Jordan Canonical Form

Theory and Practice

Synthesis Lectures on Mathematics and Statistics

Editor
Steven G. Krantz, *Washington University, St. Louis*

Jordan Canonical Form: Theory and Practice

Steven H. Weintraub

ISBN: 978-3-031-01270-9   paperback
ISBN: 978-3-031-02398-9   ebook

DOI 10.1007/978-3-031-02398-9

A Publication in the Springer series
SYNTHESIS LECTURES ON MATHEMATICS AND STATISTICS

Lecture #6
Series Editor: Steven G. Krantz, *Washington University, St. Louis*
Series ISSN
Synthesis Lectures on Mathematics and Statistics
Print 1938-1743   Electronic 1938-1751

Jordan Canonical Form

Theory and Practice

Steven H. Weintraub
Lehigh University

SYNTHESIS LECTURES ON MATHEMATICS AND STATISTICS #6

ABSTRACT

Jordan Canonical Form (*JCF*) is one of the most important, and useful, concepts in linear algebra. The *JCF* of a linear transformation, or of a matrix, encodes all of the structural information about that linear transformation, or matrix. This book is a careful development of *JCF*. After beginning with background material, we introduce Jordan Canonical Form and related notions: eigenvalues, (generalized) eigenvectors, and the characteristic and minimum polynomials. We decide the question of diagonalizability, and prove the Cayley–Hamilton theorem. Then we present a careful and complete proof of the fundamental theorem: *Let V be a finite-dimensional vector space over the field of complex numbers* $\mathbf{C}$, *and let* $T : V \longrightarrow V$ *be a linear transformation. Then T has a Jordan Canonical Form.* This theorem has an equivalent statement in terms of matrices: *Let A be a square matrix with complex entries. Then A is similar to a matrix J in Jordan Canonical Form, i.e., there is an invertible matrix P and a matrix J in Jordan Canonical Form with* $A = PJP^{-1}$. We further present an algorithm to find P and J, assuming that one can factor the characteristic polynomial of A. In developing this algorithm we introduce the eigenstructure picture (*ESP*) of a matrix, a pictorial representation that makes *JCF* clear. The *ESP* of A determines J, and a refinement, the labelled eigenstructure picture (ℓESP) of A, determines P as well. We illustrate this algorithm with copious examples, and provide numerous exercises for the reader.

KEYWORDS

Jordan Canonical Form, characteristic polynomial, minimum polynomial, eigenvalues, eigenvectors, generalized eigenvectors, diagonalizability, Cayley–Hamilton theorem, eigenstructure picture

To my brother Jeffrey and my sister Sharon

Contents

Preface

Jordan Canonical Form (*JCF*) is one of the most important, and useful, concepts in linear algebra. The *JCF* of a linear transformation, or of a matrix, encodes all of the structural information about that linear transformation, or matrix. This book is a careful development of *JCF*.

In Chapter 1 of this book we present necessary background material. We expect that most, though not all, of the material in this chapter will be familiar to the reader.

In Chapter 2 we define Jordan Canonical Form and prove the following fundamental theorem: *Let V be a finite-dimensional vector space over the field of complex numbers $\mathbb{C}$, and let $\mathcal{T} : V \longrightarrow V$ be a linear transformation. Then $\mathcal{T}$ has a Jordan Canonical Form.* This theorem has an equivalent statement in terms of matrices: *Let A be a square matrix with complex entries. Then A is similar to a matrix J in Jordan Canonical Form.* Along the way to the proof we introduce eigenvalues and (generalized) eigenvectors, and the characteristic and minimum polynomials of a linear transformation (or matrix), all of which play a key role. We also examine the special case of diagonalizability and prove the Cayley–Hamilton theorem.

The main result of Chapter 2 may be restated as: *Let A be a square matrix with complex entries. Then there is an invertible matrix P and a matrix J in Jordan Canonical Form with $A = PJP^{-1}$.* In Chapter 3 we present an algorithm to find P and J, assuming that one can factor the characteristic polynomial of A. In developing this algorithm we introduce the idea of the eigenspace picture (*ESP*) of A, which determines J, and a refinement, the labelled eigenspace picture (ℓESP) of A, which determines P as well. We illustrate this algorithm with copious examples.

Our numbering system in this text is fairly standard. Theorem 1.2.3 is the third numbered result in Section 2 of Chapter 1.

We provide many exercises for the reader to gain facility in applying these concepts and in particular in finding the *JCF* of matrices. As is customary in texts, we provide answers to the odd-numbered exercises here. *Instructors* may contact me at shw2@lehigh.edu and I will supply answers to all of the exercises.

Steven H. Weintraub
Department of Mathematics, Lehigh University
Bethlehem, PA 18015 USA
July 2009

CHAPTER 1

Fundamentals on Vector Spaces and Linear Transformations

Throughout this book, all vector spaces will be assumed to be finite-dimensional vector spaces over the field $\mathbb{C}$ of complex numbers.

We use, without further ado, the *Fundamental Theorem of Algebra*: *Let* $f(x) = a_n x^n + a_{n-1} x^{n-1} + \ldots + a_1 x + a_0$ *be a polynomial of degree n with complex coefficients. Then* $f(x) = a_n(x - r_1)(x - r_2) \cdots (x - r_n)$ *for some complex numbers* $r_1, r_2, \ldots, r_n$ *(the roots of the polynomial).*

1.1 BASES AND COORDINATES

In this section, we discuss some of the basic facts on bases for vector spaces, and on coordinates for vectors.

First, we recall the definition of a basis.

Definition 1.1.1. Let V be a vector space and let $\mathcal{B} = \{v_1, \ldots, v_n\}$ be a set of vectors of V. Then $\mathcal{B}$ is a *basis* of V if it satisfies the following two conditions:

(1) $\mathcal{B}$ spans V, i.e., any vector v in V can be written as $v = c_1 v_1 + \ldots + c_n v_n$ for some scalars $c_1, \ldots, c_n$; and

(2) $\mathcal{B}$ is linearly independent, i.e., the only solution of $0 = c_1 v_1 + \ldots + c_n v_n$ is the trivial solution $c_1 = \ldots = c_n = 0$. ◇

Lemma 1.1.2. *Let V be a vector space and let $\mathcal{B} = \{v_1, \ldots, v_n\}$ be a set of vectors in V. Then $\mathcal{B}$ is a basis of V if and only if any vector v in V can be written as $v = c_1 v_1 + \ldots + c_n v_n$ in a unique way.*

Proof. First suppose that any vector v in V can be written as $v = c_1 v_1 + \ldots + c_n v_n$ in a unique way. Then, in particular, it can be written as $v = c_1 v_1 + \ldots + c_n v_n$, so $\mathcal{B}$ spans V. Also, we certainly have $0 = 0 v_1 + \ldots + 0 v_n$, so if this is the only way to write the 0 vector, when we write $0 = c_1 v_1 + \ldots + c_n v_n$ we must have $c_1 = \ldots = c_n = 0$.

Conversely, let $\mathcal{B}$ be a basis of V. Then, since $\mathcal{B}$ spans V, we can certainly write any vector v in V in the form $v = c_1 v_1 + \ldots + c_n v_n$. Thus it remains to show that this can only be done in one way. So suppose that we also have $v = c_1' v_1 + \ldots + c_n' v_n$. Then we have $0 = v - v = (c_1 - c_1')v_1 + \ldots + (c_n - c_n')v_n$. By linear independence, we must have $c_1 - c_1' = \ldots = c_n - c_n' = 0$, i.e., $c_1' = c_1, \ldots, c_n' = c_n$, and so we have uniqueness. □

This lemma leads to the following definition.

Definition 1.1.3. Let V be a vector space and let $\mathcal{B} = \{v_1, \ldots, v_n\}$ be a basis of V. Let v be a vector in V and write $v = c_1 v_1 + \ldots + c_n v_n$. Then the vector

$$[v]_{\mathcal{B}} = \begin{bmatrix} c_1 \\ \vdots \\ c_n \end{bmatrix}$$

is the *coordinate vector* of v in the basis $\mathcal{B}$. ◇

Remark 1.1.4. In particular, we may take $V = \mathbb{C}^n$ and consider the *standard basis* $\mathcal{E}_n = \{e_1, \ldots, e_n\}$ where

$$e_i = \begin{bmatrix} 0 \\ \vdots \\ 0 \\ 1 \\ 0 \\ \vdots \end{bmatrix},$$

with 1 in the ith position, and 0 elsewhere. (We will often abbreviate $\mathcal{E}_n$ by $\mathcal{E}$ when there is no possibility of confusion.)

Then, if

$$v = \begin{bmatrix} c_1 \\ c_2 \\ \vdots \\ c_{n-1} \\ c_n \end{bmatrix},$$

we see that

$$v = c_1 \begin{bmatrix} 1 \\ 0 \\ \vdots \\ 0 \\ 0 \end{bmatrix} + \ldots + c_n \begin{bmatrix} 0 \\ 0 \\ \vdots \\ 0 \\ 1 \end{bmatrix} = c_1 e_1 + \ldots + c_n e_n,$$

so we then see that

$$[v]_{\mathcal{E}_n} = \begin{bmatrix} c_1 \\ c_2 \\ \vdots \\ c_{n-1} \\ c_n \end{bmatrix}.$$

(In other words, a vector in $\mathbb{C}^n$ "looks like" itself in the standard basis.) ◇

We have the following two properties of coordinates.

Lemma 1.1.5. *Let V be a vector space and let $\mathcal{B} = \{v_1, \ldots, v_n\}$ be any basis of V.*
(1) If $v = 0$ is the 0 vector in V, then $[v]_\mathcal{B} = 0$ is the 0 vector in $\mathbb{C}^n$.
(2) If $v = v_i$, then $[v]_\mathcal{B} = e_i$.

Proof. We leave this as an exercise for the reader.   □

Proposition 1.1.6. *Let V be a vector space and let $\mathcal{B} = \{v_1, \ldots, v_n\}$ be any basis of V.*
(1) For any vectors v_1 and v_2 in V, $[v_1 + v_2]_\mathcal{B} = [v_1]_\mathcal{B} + [v_2]_\mathcal{B}$.
(2) For any vector v_1 in V and any scalar c, $[cv_1]_\mathcal{B} = c[v_1]_\mathcal{B}$.
(3) For any vector w in $\mathbb{C}^n$, there is a unique vector v in V with $[v]_\mathcal{B} = w$.

Proof. We leave this as an exercise for the reader.   □

What is essential to us is the ability to compare coordinates in different bases. To that end, we have the following theorem.

Theorem 1.1.7. *Let V be a vector space, and let $\mathcal{B} = \{v_1, \ldots, v_n\}$ and $\mathcal{C} = \{w_1, \ldots, w_n\}$ be two bases of V. Then there is a unique matrix $P_{\mathcal{C} \leftarrow \mathcal{B}}$ with the property that, for any vector v in V,*

$$[v]_\mathcal{C} = P_{\mathcal{C} \leftarrow \mathcal{B}}[v]_\mathcal{B}.$$

This matrix is given by

$$P_{\mathcal{C} \leftarrow \mathcal{B}} = \left[[v_1]_\mathcal{C} \,\middle|\, [v_2]_\mathcal{C} \,\middle|\, \cdots \,\middle|\, [v_n]_\mathcal{C} \right].$$

Proof. Let us first check that the given matrix $P = P_{\mathcal{C} \leftarrow \mathcal{B}}$ has the desired property. To this end, let us first consider $v = v_i$ for some i. Then, as we have seen in Lemma 1.1.5 (2), $[v]_\mathcal{B} = e_i$. By the properties of matrix multiplication, Pe_i is the ith column of P, which by the definition of P is $[v_i]_\mathcal{C}$. Thus in this case, we indeed have $[v]_\mathcal{C} = P[v]_\mathcal{C}$, as claimed. Now consider a general vector v. Then on the one hand, by the definition of coordinates, if $v = c_1 v_1 + \ldots + c_n v_n$ then $[v]_\mathcal{B} = u = \begin{bmatrix} c_1 \\ \vdots \\ c_n \end{bmatrix}$.

By Proposition 1.1.6 (1) and (2), we then have that $[v]_\mathcal{C} = c_1[v]_\mathcal{C} + \ldots + c_n[v]_\mathcal{C}$, while on the other hand, by the definition of matrix multiplication, this is precisely $Pu = P[v]_\mathcal{B}$. Thus we have $[v]_\mathcal{C} = P[v]_\mathcal{C}$ for every vector.

It remains only to show that P is the unique matrix with this property. Thus, suppose we have another matrix P' with this property. To this end, let u be any vector in $\mathbb{C}^n$. Then, by Proposition 1.1.6 (3), there is a vector v in V with $[v]_\mathcal{B} = u$. Then we have $[v]_\mathcal{C} = Pu$ and $[v]_\mathcal{C} = P'u$, so $P'u = Pu$. Since u is arbitrary, the only way this can happen is if $P' = P$.   □

Again, this theorem leads to a definition.

Definition 1.1.8. The matrix $P_{C\leftarrow B}$ is the *change-of-basis matrix* from the basis B to the basis C. $\diamond$

The change-of-basis matrix has the following properties.

Proposition 1.1.9. *Let V be a vector space and let $B, C,$ and D be any three bases of V. Then:*
(1) $P_{B\leftarrow B} = I$, the identity matrix.
(2) $P_{D\leftarrow B} = P_{D\leftarrow C} P_{C\leftarrow B}$.
(3) $P_{C\leftarrow B}$ is invertible and $(P_{C\leftarrow B})^{-1} = P_{B\leftarrow C}$.

Proof. (1) Certainly $[v]_B = I[v]_B$ for every vector v in V, so by the uniqueness of the change-of-basis matrix, we must have $P_{B\leftarrow B} = I$.

(2) Using the fact that matrix multiplication is associative, we have

$$(P_{D\leftarrow C} P_{C\leftarrow B})[v]_B = P_{D\leftarrow C}(P_{C\leftarrow B}[v]_B) = P_{D\leftarrow C}[v]_C = [v]_D$$

for every vector v in V, so, again by the uniqueness of the change-of-basis matrix, we must have $P_{D\leftarrow B} = P_{D\leftarrow C} P_{C\leftarrow B}$.

(3) Setting $D = B$ and using parts (1) and (2) we have that $I = P_{B\leftarrow B} = P_{B\leftarrow C} P_{C\leftarrow B}$. Similarly, we have that $I = P_{C\leftarrow C} = P_{C\leftarrow B} P_{B\leftarrow C}$, and hence these two matrices are inverses of each other. $\square$

We have just seen, in Proposition 1.1.9 (3), that every change-of-basis matrix is invertible. In fact, every invertible matrix occurs as a change-of-basis matrix, as we see from the next Proposition.

Proposition 1.1.10. *Let P be any invertible n-by-n matrix. Then there are bases B and C of $V = \mathbb{C}^n$ with $P_{C\leftarrow B} = P$.*

Proof. Write P as

$$P = \left[p_1 \mid p_2 \mid \cdots \mid p_n \right].$$

Let $B = \{p_1, p_2, \ldots, p_n\}$. Since P is invertible, B is a basis of $\mathbb{C}^n$. Let $C = \mathcal{E}$ be the standard basis of $V = \mathbb{C}^n$. We saw in Example 1.1.4 that $[v]_{\mathcal{E}} = v$ for any vector v in V. In particular, this is true for $v = p_i$, for each i. But then, by Theorem 1.1.7,

$$P_{\mathcal{E}\leftarrow B} = \left[[p_1]_{\mathcal{E}} \mid [p_2]_{\mathcal{E}} \mid \cdots \mid [p_n]_{\mathcal{E}} \right] = \left[p_1 \mid p_2 \mid \cdots \mid p_n \right] = P,$$

as claimed. $\square$

1.2   LINEAR TRANSFORMATIONS AND MATRICES

In this section, we introduce linear transformations and their matrices.

First, we recall the definition of a linear transformation.

Definition 1.2.1. Let V and W be vector spaces. A *linear transformation* $\mathcal{T}$ is a function $\mathcal{T} : V \longrightarrow W$ satisfying the following two properties:

(1) $\mathcal{T}(v_1 + v_2) = \mathcal{T}(v_1) + \mathcal{T}(v_2)$ for any two vectors v_1 and v_2 in V; and

(2) $\mathcal{T}(cv_1) = c(v_1)$ for any vector v_1 in V and any scalar c.   ◇

A linear transformation has the following important property.

Lemma 1.2.2. *Let V and W be vector spaces, and let $\mathcal{B} = \{v_1, \ldots, v_n\}$ be any basis of V. Then a linear transformation $\mathcal{T} : V \longrightarrow W$ is determined by $\{\mathcal{T}(v_1), \ldots, \mathcal{T}(v_n)\}$, which may be specified arbitrarily.*

Proof. Suppose we are given $\mathcal{T}(v_i) = w_i$ for each i. Choose any vector v in V. Then, by Lemma 1.1.2, v can be written uniquely as $v = \sum c_i v_i$, and then, by the properties of a linear transformation, $\mathcal{T}(v) = \mathcal{T}(\sum c_i v_i) = \sum c_i \mathcal{T}(v_i) = \sum c_i w_i$ is determined.

On the other hand, suppose we choose any $\{w_1, \ldots, w_n\}$. We define $\mathcal{T}$ as follows: First we set $\mathcal{T}(v_i) = w_i$ for each i. Then any vector v can be written as $v = \sum c_i v_i$, so we set $\mathcal{T}(v) = \sum c_i w_i$. This gives us a well-defined function as there is only one such way to write v. It remains to check that the function $\mathcal{T}$ defined in this way is a linear transformation, and we leave that to the reader.   □

Here is one of the most important ways of constructing linear transformations.

Lemma 1.2.3. *Let A be an m-by-n matrix and let $\mathcal{T}_A : \mathbb{C}^n \to \mathbb{C}^m$ be defined by $\mathcal{T}_A(v) = Av$. Then $\mathcal{T}_A(v)$ is a linear transformation.*

Proof. This follows directly from properties of matrix multiplication: $\mathcal{T}_A(v_1 + v_2) = A(v_1 + v_2) = Av_1 + Av_2 = \mathcal{T}_A(v_1) + \mathcal{T}_A(v_2)$ and $\mathcal{T}_A(cv_1) = A(cv_1) = c(Av_1) = c\mathcal{T}_A(v_1)$.   □

In fact, once we choose bases, any linear transformation between finite-dimensional vector spaces is given by multiplication by a matrix.

Theorem 1.2.4. *Let V and W be vector spaces and let $\mathcal{T} : V \longrightarrow W$ be a linear transformation. Let $\mathcal{B} = \{v_1, \ldots, v_n\}$ be a basis of V and let $\mathcal{B} = \{w_1, \ldots, w_m\}$ be a basis of W. Then there is a unique matrix $[\mathcal{T}]_{\mathcal{C} \leftarrow \mathcal{B}}$ such that, for any vector v in V,*

$$[\mathcal{T}(v)]_{\mathcal{C}} = [\mathcal{T}]_{\mathcal{C} \leftarrow \mathcal{B}}[v]_{\mathcal{B}}.$$

Furthermore, the matrix $[\mathcal{T}]_{\mathcal{C} \leftarrow \mathcal{B}}$ is given by

$$[\mathcal{T}]_{\mathcal{C} \leftarrow \mathcal{B}} = \left[[\mathcal{T}(v_1)]_{\mathcal{C}} \,\middle|\, [\mathcal{T}(v_2)]_{\mathcal{C}} \,\middle|\, \cdots \,\middle|\, [\mathcal{T}(v_n)]_{\mathcal{C}} \right].$$

Proof. Let A be the given matrix. Consider any element v_i of the basis $\mathcal{B}$. Then $[v_i]_{\mathcal{B}} = e_i$, and so $\mathcal{T}_A([v_i]_{\mathcal{B}}) = \mathcal{T}_A([e_i]) = Ae_i$ is the ith column of A, which by the definition of the matrix A is $[\mathcal{T}(v_i)]_{\mathcal{C}}$. Thus, we have verified that $[\mathcal{T}(v_i)]_{\mathcal{C}} = [\mathcal{T}]_{\mathcal{C} \leftarrow \mathcal{B}}[v_i]_{\mathcal{B}}$ for each element of the basis $\mathcal{B}$. Since this is true on a basis, it must be true for every element v of V, again using the fact that we can write every element v of V uniquely as $v = c_1 v_1 + \ldots + c_n v_n$. (Compare the proof of Theorem 1.1.7.)

Thus the matrix A has the property we have claimed. It remains to show that it is the unique matrix with this property. Suppose we had another matrix A' with the same property. Then we would have $A[v]_B = [T(v)]_C = A'[v]_B$ for every v in V. Since for any vector u in $\mathbb{C}^n$ there is a vector v with $[v]_B = u$, we see that $Au = A'u$ for every vector u in $\mathbb{C}^n$, so we must have $A' = A$. (Again compare the proof of Theorem 1.1.7.)   □

Similarly, this theorem leads to a definition.

Definition 1.2.5. Let V and W be vector spaces and let $T : V \longrightarrow V$ be a linear transformation. Let $B = \{v_1, \ldots, v_n\}$ be a basis of V and let $B = \{w_1, \ldots, w_m\}$ be a basis of W. Let $[T]_{C \leftarrow B}$ be the matrix defined in Theorem 1.2.4. Then $[T]_{C \leftarrow B}$ is the *matrix of the linear transformation* T from the basis B to the basis C.   ◇

Remark 1.2.6. In particular, we may take $V = \mathbb{C}^n$ and $W = \mathbb{C}^m$, and consider the standard bases $\mathcal{E}_n$ of V and $\mathcal{E}_m$ of W. Let A be an m-by-n square matrix and write $A = \left[a_1 \,\middle|\, a_2 \,\middle|\, \ldots \,\middle|\, a_n \right]$. If T_A is the linear transformation given by $T_A(v) = Av$, then

$$
\begin{aligned}
[T_A]_{\mathcal{E}_m \leftarrow \mathcal{E}_n} &= \left[[T_A(e_1)]_{\mathcal{E}_m} \,\middle|\, [T_A(e_2)]_{\mathcal{E}_m} \,\middle|\, \ldots \,\middle|\, [T_A(e_n)]_{\mathcal{E}_m} \right] \\
&= \left[[Ae_1]_{\mathcal{E}_m} \,\middle|\, [Ae_2]_{\mathcal{E}_m} \,\middle|\, \ldots \,\middle|\, [Ae_n]_{\mathcal{E}_m} \right] \\
&= \left[[a_1]_{\mathcal{E}_m} \,\middle|\, [a_2]_{\mathcal{E}_m} \,\middle|\, \ldots \,\middle|\, [a_n]_{\mathcal{E}_m} \right] \\
&= \left[a_1 \,\middle|\, a_2 \,\middle|\, \ldots \,\middle|\, a_n \right] \\
&= A.
\end{aligned}
$$

(In other words, the linear transformation given by multiplication by a matrix "looks like" that same matrix from one standard basis to another.)   ◇

There is a very important special case of Definition 1.2.5, when T is a linear transformation from the vector space V to itself. In this case, we may choose the basis C to just be the basis B. We can also simplify the notation, as in this case, we do not need to record the same basis twice.

Definition 1.2.7. Let V be a vector space and let $T : V \longrightarrow V$ be a linear transformation. Let $B = \{v_1, \ldots, v_n\}$ be a basis of V. Let $[T]_B = [T]_{B \leftarrow B}$ be the matrix defined in Theorem 1.2.4. Then $[T]_B$ is the *matrix of the linear transformation* T in the basis B.   ◇

What is essential to us is the ability to compare matrices of linear transformations in different bases. To that end, we have the following theorem. (Compare Theorem 1.1.7.)

Theorem 1.2.8. *Let V be a vector space, and let $B = \{v_1, \ldots, v_n\}$ and $C = \{w_1, \ldots, w_n\}$ be any two bases of V. Let $P_{C \leftarrow B}$ be the change-of-basis matrix form B to C. Then, for any linear transformation*

$$T : V \longrightarrow V,$$

$$
\begin{aligned}
[T]_{\mathcal{C}} &= P_{\mathcal{C} \leftarrow \mathcal{B}} [T]_{\mathcal{B}} P_{\mathcal{B} \leftarrow \mathcal{C}} \\
&= P_{\mathcal{C} \leftarrow \mathcal{B}} [T]_{\mathcal{B}} (P_{\mathcal{C} \leftarrow \mathcal{B}})^{-1} \\
&= (P_{\mathcal{B} \leftarrow \mathcal{C}})^{-1} [T]_{\mathcal{B}} P_{\mathcal{B} \leftarrow \mathcal{C}}.
\end{aligned}
$$

Proof. Let v be any vector in V. We compute:

$$
\begin{aligned}
(P_{\mathcal{C} \leftarrow \mathcal{B}} [T]_{\mathcal{B}} P_{\mathcal{B} \leftarrow \mathcal{C}})([v]_{\mathcal{C}}) &= (P_{\mathcal{C} \leftarrow \mathcal{B}} [T]_{\mathcal{B}})(P_{\mathcal{B} \leftarrow \mathcal{C}} [v]_{\mathcal{C}}) \\
&= (P_{\mathcal{C} \leftarrow \mathcal{B}} [T]_{\mathcal{B}})([v]_{\mathcal{B}}) \\
&= P_{\mathcal{C} \leftarrow \mathcal{B}} ([T]_{\mathcal{B}} [v]_{\mathcal{B}}) \\
&= P_{\mathcal{C} \leftarrow \mathcal{B}} [T(v)]_{\mathcal{B}} \\
&= [T(v)]_{\mathcal{C}}.
\end{aligned}
$$

But, by definition, $[T]_{\mathcal{C}}$ is the unique matrix with the property that $[T]_{\mathcal{C}}([v]_{\mathcal{C}}) = [T(v)]_{\mathcal{C}}$ for every v in V. Thus these two matrices must be the same, yielding the first equality in the statement of the theorem. The second and third equalities then follow from Proposition 1.1.9 (3). $\square$

Again, this theorem leads to a definition.

Definition 1.2.9. Two square matrices A and B are *similar* if there is an invertible matrix P with $A = PBP^{-1}$. A is said to be obtained from B by *conjugation* by P. The matrix P is called an *intertwining matrix* of A and B, or is said to *intertwine* A and B. $\diamond$

Remark 1.2.10. Note that if $A = PBP^{-1}$, so that A is obtained from B by conjugation by P, or equivalently that P intertwines A and B, then $B = P^{-1}AP$, so that B is obtained from A by conjugation by P^{-1}, or equivalently that P^{-1} intertwines B and A. $\diamond$

Theorem 1.2.8 tells us that if A and B are matrices of the same linear transformation T with respect to two bases of V, then A and B are similar. In fact, any pair of similar matrices arises in this way. (Compare Proposition 1.1.10.)

Proposition 1.2.11. *Let A and B be any two similar n–by–n matrices. Then there is a linear transformation $T : \mathbb{C}^n \longrightarrow \mathbb{C}^n$ and bases $\mathcal{B}$ and $\mathcal{C}$ of $\mathbb{C}^n$ such that $[T]_{\mathcal{B}} = A$ and $[T]_{\mathcal{C}} = B$.*

Proof. Let $A = PBP^{-1}$, or, equivalently, $B = P^{-1}AP$. Let T be the linear transformation $T = T_A$. Let $\mathcal{B}$ be the basis of $\mathbb{C}^n$ whose elements are the columns of P^{-1}, and let $\mathcal{C} = \mathcal{E}$ be the standard basis of $\mathbb{C}^n$. Then, by Remark 1.2.6, $[T]_{\mathcal{E}} = A$, and then, by Theorem 1.2.8, $[T]_{\mathcal{B}} = P^{-1}AP = B$. $\square$

Remark 1.2.12. This point is so important that it is worth specially emphasizing: Two matrices A and B are similar if and only if they are the matrices of the *same* linear transformation T in two bases. $\diamond$

1.3   SOME SPECIAL MATRICES

In this section, we present a small menagerie of matrices of special forms. These matrices are simpler than general matrices, and will be important for us to consider in the sequel.

All matrices we consider here are square matrices.

Definition 1.3.1. A matrix A is a *scalar matrix* if it is a scalar multiple of the identity matrix, $A = cI$. Equivalently, a scalar matrix is a matrix of the form

$$\begin{bmatrix} c & & & & & \\ & c & & & 0 & \\ & & c & & & \\ & & & \ddots & & \\ & 0 & & & c & \\ & & & & & c \end{bmatrix}.$$

◇

Definition 1.3.2. A matrix A is a *diagonal matrix* if all of its entries that do not lie on the diagonal are 0. Equivalently, a diagonal matrix is a matrix of the form

$$\begin{bmatrix} d_1 & & & & & \\ & d_2 & & 0 & & \\ & & d_3 & & & \\ & & & \ddots & & \\ & 0 & & & d_{n-1} & \\ & & & & & d_n \end{bmatrix}.$$

◇

Definition 1.3.3. A matrix A is an *upper triangular matrix* if all of its entries that lie below the diagonal are 0. Equivalently, an upper triangular matrix is a matrix of the form

$$\begin{bmatrix} a_{1,1} & & & & & \\ & a_{2,2} & & * & & \\ & & a_{3,3} & & & \\ & & & \ddots & & \\ & 0 & & & a_{n-1,n-1} & \\ & & & & & a_{n,n} \end{bmatrix}.$$

(Here we introduce the notation that $*$ stands for a matrix entry, or group of entries, that may be arbitrary.)

◇

Definition 1.3.4. A matrix A is an *upper triangular matrix with constant diagonal* if it is an upper triangular matrix with all of its diagonal entries equal. Equivalently, an upper triangular matrix with constant diagonal is a matrix of the form

$$\begin{bmatrix} a & & & & & \\ & a & & & * & \\ & & a & & & \\ & & & \ddots & & \\ & 0 & & & a & \\ & & & & & a \end{bmatrix}.$$

$\diamond$

Definition 1.3.5. A matrix A is a *block diagonal matrix* if it is a matrix of the form

$$\begin{bmatrix} A_1 & & & & & \\ & A_2 & & 0 & & \\ & & A_3 & & & \\ & & & \ddots & & \\ & 0 & & & A_{m-1} & \\ & & & & & A_m \end{bmatrix}$$

for some matrices $A_1, A_2, \ldots, A_m$.

$\diamond$

Definition 1.3.6. A matrix A is a *block diagonal matrix with blocks scalar matrices*, or a *BDBSM matrix* if it is a block diagonal matrix with each block A_i a scalar matrix.

$\diamond$

Remark 1.3.7. Of course, a *BDBSM* matrix is itself a diagonal matrix.

$\diamond$

Definition 1.3.8. A matrix A is a *block diagonal matrix whose blocks are upper triangular with constant diagonal*, or a *BDBUTCD matrix* if it is a block diagonal matrix with each block A_i an upper triangular matrix with constant diagonal.

$\diamond$

Remark 1.3.9. Of course, a *BDBUTCD* matrix is itself an upper triangular matrix.

$\diamond$

Definition 1.3.10. A matrix A is a *block upper triangular matrix* if it is a matrix of the form

$$\begin{bmatrix} A_1 & & & & & \\ & A_2 & & * & & \\ & & A_3 & & & \\ & & & \ddots & & \\ & 0 & & & A_{m-1} & \\ & & & & & A_m \end{bmatrix}$$

for some matrices $A_1, A_2, \ldots, A_m$.

$\diamond$

You could easily (or perhaps not so easily) imagine why we would want to study the above kinds of matrices, as they are simpler than arbitrary matrices and occur more or less naturally. Our last two kinds of matrices are ones you would certainly not come up with offhand, but turn out to be vitally important ones. Indeed, these are precisely the matrices we are heading towards. It will take us a lot of work to get there, but it is easy for us to simply define them now and concern ourselves with their meaning later.

Definition 1.3.11. A *Jordan block* J is a matrix of the following form:

$$
J = \begin{bmatrix}
a & 1 & & & & \\
 & a & 1 & & 0 & \\
 & & a & 1 & & \\
 & & & \ddots & \ddots & \\
 & 0 & & & a & 1 \\
 & & & & & a
\end{bmatrix}.
$$

In other words, a Jordan block is an upper triangular matrix of a very special form: All of the diagonal entries are equal to each other; all of the entries immediately above the diagonal are equal to 1; and all of the other entries are equal to 0.

Definition 1.3.12. A matrix is in *Jordan Canonical Form (JCF)*, or is a *Jordan matrix*, if it is a matrix of the form

$$
\begin{bmatrix}
J_1 & & & & & \\
 & J_2 & & 0 & & \\
 & & J_3 & & & \\
 & & & \ddots & & \\
 & 0 & & & J_{m-1} & \\
 & & & & & J_m
\end{bmatrix},
$$

where each block J_i is a Jordan block.

Remark 1.3.13. We see that a matrix in *JCF* is a very particular kind of *BDBUTCD* matrix.

We now define a general class of matrices, and we examine some examples of matrices in that class.

Definition 1.3.14. A matrix A such that some power $A^k = 0$ is *nilpotent*. If A is nilpotent, the smallest positive integer k with $A^k = 0$ is the *index of nilpotency* of A.

Definition 1.3.15. A matrix A is *strictly upper triangular* if it is upper triangular with all of its diagonal entries equal to 0.

Lemma 1.3.16. *Let A be an n-by-n strictly upper triangular matrix. Then A is nilpotent with index of nilpotency at most n.*

Proof. We give two proofs. The first proof is short but not very enlightening. The second proof is longer but more enlightening.

First proof: Direct computation shows that for any strictly upper triangular matrix A, $A^n = 0$. (A has all of its diagonal entries 0; A^2 has all of its diagonal entries and all of its entries immediately above the diagonal 0; A^3 has all of its entries on, one space above, and two spaces above the diagonal 0; etc.)

Second proof: Let $\mathcal{E} = \{e_1, \ldots, e_n\}$ be the standard basis of $\mathbb{C}^n$ and define subspaces $V_0, \ldots,$ V_n of $\mathbb{C}^n$ as follows: $V_0 = \{0\}$, V_1 is the subspace spanned by the vector e_1, V_2 is the subspace spanned by the vectors e_1 and e_2, etc., (so that, in particular, $V_n = \mathbb{C}^n$). If A is strictly upper triangular, then $AV_i \subseteq V_{i-1}$ for every $i \geq 1$ (and in particular $AV_1 = \{0\}$). But then $A^2 V_i \subseteq V_{i-2}$ for every $i \geq 2$, and $A^2 V_2 = A^2 V_1 = \{0\}$. Proceeding in this way we see that $A_n \mathbb{C}^n = A^n V_n = \{0\}$, i.e., $A^n = 0$. □

Lemma 1.3.17. *Let J be an n-by-n Jordan block with diagonal entry a. Then $J - aI$ is nilpotent with index of nilpotency n.*

Proof. Again we present two proofs.

First proof: $A = J - aI$ is strictly upper triangular, so by Lemma 1.3.16 $(J - aI)^k = 0$ for some $k \leq n$. But direct computation shows that $(J - aI)^{k-1} \neq 0$ (it is the matrix with a 1 in the upper right-hand corner and all other entries 0), so $J - aI$ must have index of nilpotency n.

Second proof: Let $A = J - aI$. In the notation of the second proof of Lemma 1.3.16, $AV_i = V_{i-1}$ for every $i \geq 1$. Then $A^{n-1} V_n = V_1 \neq \{0\}$, so $A^{n-1} \neq 0$, but $A^n V_n = V_0 = \{0\}$, so $A^n = 0$, and A has index of nilpotency n. □

1.4   POLYNOMIALS IN $\mathcal{T}$ AND A

In this section, we consider polynomials in linear transformations $\mathcal{T}$ or matrices A. We begin with some general facts, and then examine some special cases.

All matrices we consider here are square matrices, and all polynomials we consider here have their coefficients in the field of complex numbers.

Definition 1.4.1. Let $f(x) = a_n x^n + a_{n-1} x^{n-1} + \ldots + a_1 x + a_0$ be an arbitrary polynomial.

(1) If $\mathcal{T} : V \longrightarrow V$ is a linear transformation, we let

$$f(\mathcal{T}) = a_n \mathcal{T}^n + a_{n-1} \mathcal{T}^{n-1} + a_1 \mathcal{T} + a_0 \mathcal{I}$$

where $\mathcal{T}^n$ denotes the n-fold composition of $\mathcal{T}$ with itself (i.e., $\mathcal{T}^2(v) = \mathcal{T}(\mathcal{T}(v))$, $\mathcal{T}^3(v) = \mathcal{T}(\mathcal{T}(\mathcal{T}(v)))$, for every v in V, etc.) and $\mathcal{I}$ denotes the identity transformation

$(\mathcal{I}(v) = v$ for every v in V).

(2) If A is a square matrix, we let

$$f(A) = a_n A^n + a_{n-1} A^{n-1} + a_1 A + a_0 I.$$

$\diamond$

Lemma 1.4.2. *Let $\mathcal{T} : V \longrightarrow V$ be a linear transformation. Let $\mathcal{B}$ be any basis of V, and let $A = [\mathcal{T}]_{\mathcal{B}}$. Then $f(A) = [f(\mathcal{T})]_{\mathcal{B}}$.*

Proof. Matrix multiplication is defined as it is precisely by the property that, for any two linear transformations $\mathcal{S}$ and $\mathcal{T}$ from V to V, if $A = [\mathcal{S}]_{\mathcal{B}}$, and $B = [\mathcal{T}]_{\mathcal{B}}$, then $AB = [\mathcal{ST}]_{\mathcal{B}}$. The lemma easily follows from this special case. $\square$

Corollary 1.4.3. *Let $\mathcal{T} : V \longrightarrow V$ be a linear transformation. Let $\mathcal{B}$ be any basis of V, and let $A = [\mathcal{T}]_{\mathcal{B}}$. Then $f(\mathcal{T}) = 0$ if and only if $f(A) = 0$.*

Proof. By Lemma 1.4.2, $f(A) = [f(\mathcal{T})]_{\mathcal{B}}$. But a linear transformation is the 0 linear transformation if and only if its matrix (in any basis) is the 0 matrix. $\square$

Lemma 1.4.4. *Let $f(x)$ and $g(x)$ be arbitrary polynomials, and let $h(x) = f(x)g(x)$.*

(1) For any linear transformation $\mathcal{T}$, $f(\mathcal{T})g(\mathcal{T}) = h(\mathcal{T}) = g(\mathcal{T})f(\mathcal{T})$.
(2) For any square matrix A, $f(A)g(A) = h(A) = g(A)f(A)$.

Proof. We leave this as an exercise for the reader. (It looks obvious, but there is actually something to prove, and if you try to prove it, you will notice that you use the fact that any two powers of $\mathcal{T}$, or any two powers of A, commute with each other, something that is not true for arbitrary linear transformations or matrices). $\square$

Lemma 1.4.5. *Let A and B be similar square matrices, and let P intertwine A and B. Then, for any polynomial $f(x)$, P intertwines $f(A)$ and $f(B)$. In particular:*

(1) $f(A)$ and $f(B)$ are similar; and

(2) $f(A) = 0$ if and only if $f(B) = 0$.

Proof. Certainly $I = PIP^{-1}$. We are given that $A = PBP^{-1}$. Then $A^2 = (PBP^{-1})^2 = (PBP^{-1})(PBP^{-1}) = PB(P^{-1}P)BP^{-1} = PB^2P^{-1}$, and then $A^3 = A^2A = (PB^2P^{-1})(PBP^{-1}) = PB^2(P^{-1}P)BP^{-1} = PB^3P^{-1}$, and similarly $A^k = PB^kP^{-1}$ for every positive integer k, from which the first part of the lemma easily follows.

Since we have found a matrix that intertwines $f(A)$ and $f(B)$, they are certainly similar. Furthermore, if $f(B) = 0$, then $f(A) = Pf(B)P^{-1} = P0P^{-1} = 0$, while if $f(A) = 0$, then $f(B) = P^{-1}f(A)P = P^{-1}0P = 0$. $\square$

Lemma 1.4.6. *Let $f(x)$ be an arbitrary polynomial.*

(1) If $A = cI$ is a scalar matrix, then $f(A)$ is the scalar matrix $f(A) = f(c)I$.

(2) If A is the diagonal matrix with entries $d_1, d_2, \ldots, d_n$, then $f(A)$ is the diagonal matrix with entries $f(d_1), f(d_2), \ldots, f(d_n)$.

(3) If A is an upper triangular matrix with entries $a_{1,1}, a_{2,2}, \ldots, a_{n,n}$, then $f(A)$ is an upper triangular matrix with entries $f(a_{1,1}), f(a_{2,2}), \ldots, f(a_{n,n})$.

Proof. Direct computation. $\square$

Lemma 1.4.7. *Let $f(x)$ be an arbitrary polynomial.*

(1) If A is a block diagonal matrix with diagonal blocks $A_1, A_2, \ldots, A_m$, then $f(A)$ is the block diagonal matrix with entries $f(A_1), f(A_2), \ldots, f(A_m)$.

(3) If A is a block upper triangular matrix with diagonal blocks $A_{1,1}, A_{2,2}, \ldots, A_{m,m}$, then $f(A)$ is a block upper triangular matrix with diagonal blocks $f(A_{1,1}), f(A_{2,2}), \ldots, f(A_{m,m})$.

Proof. Direct computation. $\square$

Note that in Lemma 1.4.6 (3) we do not say anything about the off-diagonal entries of $f(A)$, and that in Lemma 1.4.7 (2) we do not say anything about the off-diagonal blocks of $f(A)$.

1.5   SUBSPACES, COMPLEMENTS, AND INVARIANT SUBSPACES

We begin this section by considering the notion of the complement of a subspace. Afterwards, we introduce a linear transformation and reconsider the situation.

Definition 1.5.1. Let V be a vector space, and let W_1 be a subspace of V. A subspace W_2 of V is a *complementary subspace* of W_1, or simply a *complement* of W_1, if every vector v in V can be written uniquely as $v = w_1 + w_2$ for some vectors w_1 in W_1 and w_2 in W_2. In this situation, V is the *direct sum* of W_1 and W_2, $V = W_1 \oplus W_2$. $\diamond$

Note that this definition is symmetric: If W_2 is a complement of W_1, then W_1 is a complement of W_2.

Here is a criterion for a pair of subspaces to be mutually complementary.

Lemma 1.5.2. *Let W_1 and W_2 be subspaces of V. Let $\mathcal{B}_1$ be any basis of W_1 and $\mathcal{B}_2$ be any basis of W_2. Let $\mathcal{B} = \mathcal{B}_1 \cup \mathcal{B}_2$. Then W_1 and W_2 are mutually complementary subspaces of V if and only if $\mathcal{B}$ is a basis of V.*

Proof. Let $\mathcal{B}_1$ and $\mathcal{B}_2$ be the bases $\mathcal{B}_1 = \{v_{1,1}, \ldots, v_{1,i_1}\}$ and $\mathcal{B}_2 = \{v_{2,1}, \ldots, v_{2,i_2}\}$. Then $\mathcal{B} = \{v_{1,1}, \ldots, v_{1,i_1}, v_{2,1}, \ldots, v_{2,i_2}\}$. First suppose that $\mathcal{B}$ is a basis of V. Then $\mathcal{B}$ spans V, so we may write any v in V as

$$
\begin{aligned}
v &= c_{1,1}v_{1,1} + \ldots + c_{1,i_1}v_{1,i_1} + c_{2,1}v_{2,1} + \ldots + c_{2,i_2}v_{2,i_2} \\
&= (c_{1,1}v_{1,1} + \ldots + c_{1,i_1}v_{1,i_1}) + (c_{2,1}v_{2,1} + \ldots + c_{2,i_2}v_{2,i_2}) \\
&= w_1 + w_2
\end{aligned}
$$

where $w_1 = c_{1,1}v_{1,1} + \ldots + c_{1,i_1}v_{1,i_1}$ is in W_1 and $w_2 = c_{2,1}v_{2,1} + \ldots + c_{2,i_2}v_{2,i_2}$ is in W_2.

Now we must show that this way of writing v is unique. So suppose also $v = w_1' + w_2'$ with $w_1' = c_{1,1}'v_{1,1} + \ldots + c_{1,i_1}'v_{1,i_1}$ and $w_2' = c_{2,1}'v_{2,1} + \ldots + c_{2,i_2}'v_{2,i_2}$. Then $0 = v - v = (c_{1,1} - c_{1,1}')v_{1,1} + \ldots + (c_{1,i_1} - c_{1,i_1}')v_{1,i_1} + (c_{2,1} - c_{2,1}')v_{2,1} + \ldots + (c_{2,i_2} - c_{2,i_2}')v_{1,i_2}$. But $\mathcal{B}$ is linearly independent, so we must have $c_{1,1} - c_{1,1}' = 0$, ..., $c_{1,i_1} - c_{1,i_1}' = 0$, ..., $c_{2,1} - c_{2,1}' = 0$, ..., $c_{2,i_2} - c_{2,i_2}' = 0$, i.e., $c_{1,1}' = c_{1,1}$, ..., $c_{1,i_1}' = c_{1,i_1}, c_{2,1}' = c_{2,1}$, ..., $c_{2,i_2}' = c_{2,i_2}$, so $w_1' = w_1$ and $w_2' = w_2$.

We leave the proof of the converse as an exercise for the reader. $\qquad\square$

Corollary 1.5.3. *Let W_1 be any subspace of V. Then W_1 has a complement W_2.*

Proof. Let $\mathcal{B}_1 = \{v_{1,1}, \ldots, v_{1,i_1}\}$ be any basis of W_1. Then $\mathcal{B}_1$ is a linearly independent set of vectors in V, so extends to a basis $\mathcal{B} = \{v_{1,1}, \ldots, v_{1,i_1}, v_{2,1}, \ldots, v_{2,i_2}\}$ of V. Then $\mathcal{B}_2 = \{v_{2,1}, \ldots, v_{2,i_2}\}$ is a linearly independent set of vectors in V. Let W_2 be the subspace of V having $\mathcal{B}_2$ as a basis. Then W_2 is a complement of W_1. $\qquad\square$

Remark 1.5.4. If $W_1 = \{0\}$ then W_1 has the unique complement $W_2 = V$, and if $W_1 = V$ then W_1 has the unique complement $W_2 = \{0\}$. But if W_1 is a nonzero proper subspace of V, it *never* has a unique complement. For there is always more than one way to extend a basis of W_1 to a basis of V. $\diamond$

We will not only want to consider a pair of subspaces, but rather a family of subspaces, and we can readily generalize Definition 1.5.1 and Lemma 1.5.2.

Definition 1.5.5. Let V be a vector space. A set $\{W_1, \ldots, W_k\}$ of subspaces of V is a *complementary set of subspaces* of V if every vector v in V can be written uniquely as $v = w_1 + \ldots + w_k$ with w_i in W_i for each i. In this situation, V is the *direct sum* of $W_1, \ldots, W_k$, $V = W_1 \oplus \ldots \oplus W_k$.

$\diamond$

Lemma 1.5.6. *Let W_i be a subspace of V, for $i = 1, \ldots, k$. Let $\mathcal{B}_i$ be any basis of W_i, for each i. Let $\mathcal{B} = \mathcal{B}_1 \cup \ldots \cup \mathcal{B}_k$. Then $\{W_1, \ldots, W_k\}$ is a complementary set of subspaces of V if and only if $\mathcal{B}$ is a basis of V.*

Proof. We leave this as an exercise for the reader. □

Now we introduce a linear transformation and see what happens.

Definition 1.5.7. Let $\mathcal{T} : V \longrightarrow V$ be a linear transformation. A subspace W_1 of V is *invariant* under $\mathcal{T}$, or $\mathcal{T}$-*invariant*, if $\mathcal{T}(W_1) \subseteq W_1$, i.e., if $\mathcal{T}(w_1)$ is in W_1 for every vector w_1 in W_1. ◇

With the help of matrices, we can easily recognize invariant subspaces.

Lemma 1.5.8. *Let* $\mathcal{T} : V \longrightarrow V$ *be a linear transformation, where* V *is an n-dimensional vector space. Let* W *be an m-dimensional subspace of* V *and let* $\mathcal{B}_1$ *be any basis of* W. *Extend* $\mathcal{B}_1$ *to a basis* $\mathcal{B}$ *of* V. *Then* W *is a* $\mathcal{T}$-*invariant subspace of* V *if and only if*

$$Q = [\mathcal{T}]_{\mathcal{B}} = \begin{bmatrix} A & B \\ 0 & D \end{bmatrix}$$

is a block upper triangular matrix, where A *is an m-by-m block.*

Proof. Note that the condition for Q to be of this form is simply that the lower left-hand corner of Q is 0.

Let $\mathcal{B}_1 = \{v_1, \ldots, v_m\}$ and let $\mathcal{B} = \{v_1, \ldots, v_m, v_{m+1}, \ldots, v_n\}$. By Theorem 1.2.4 we know that

$$Q = \left[q_1 \,\middle|\, q_2 \,\middle|\, \cdots \,\middle|\, q_n \right] = \left[[\mathcal{T}(v_1)]_{\mathcal{B}} \,\middle|\, [\mathcal{T}(v_2)]_{\mathcal{B}} \,\middle|\, \cdots \,\middle|\, [\mathcal{T}(v_n)]_{\mathcal{B}} \right].$$

First suppose that W is $\mathcal{T}$-invariant. Then for each i between 1 and m, $\mathcal{T}(v_i)$ is in W, so for some scalars $a_{1,i}, \ldots, a_{m,i}$,

$$\mathcal{T}(v_i) = a_{1,i} v_1 + \ldots + a_{m,i} v_m$$
$$= a_{1,i} v_1 + \ldots + a_{m,i} v_m + 0 v_{m+1} + \ldots + 0 v_n$$

so

$$[\mathcal{T}(v_i)]_{\mathcal{B}} = \begin{bmatrix} a_{1,i} \\ \vdots \\ a_{m,i} \\ 0 \\ \vdots \\ 0 \end{bmatrix}$$

and Q is block upper triangular as claimed.

Conversely, suppose that Q is block upper triangular as above. Let $v = v_i$ for i between 1 and m. Then, if A has entries $\{a_{i,j}\}$, then, again by Theorem 1.2.4,

$$\mathcal{T}(v_i) = a_{1,i} v_1 + \ldots + a_{m,i} v_m + 0 v_{m+1} + \ldots + 0 v_n$$
$$= a_{1,i} v_1 + \ldots + a_{m,i} v_m$$

and so $T(v_i)$ is in W. Since this is true for every element of the basis B_1 of W, it is true for every element of W, and so $T(W) \subseteq W$, and W is T-invariant. $\square$

Corollary 1.5.9. *Let $T : V \longrightarrow V$ be a linear transformation, where V is an n-dimensional vector space. Let W_1 be an m-dimensional T-invariant subspace of V, and let W_2 be a complement of W_1. Let $B = B_1 \cup B_2$ (so that B is a basis of V). Then W_2 is a T-invariant complement of W_1 if and only if*

$$Q = [T]_B = \begin{bmatrix} A & 0 \\ 0 & D \end{bmatrix}$$

is a block diagonal matrix, where A is an m-by-m block, and D is an $(n - m)$-by-$(n - m)$ block.

Proof. The same argument in the proof of Lemma 1.5.8 that shows that W_1 is T-invariant if and only if the lower left hand corner of Q is 0 shows that W_2 is T-invariant if and only if the upper right hand corner of Q is 0. $\square$

Remark 1.5.10. Given any T-invariant subspace W_1, we can certainly find a complement W_2 (in fact, we can do this for any subspace W_1). From the matrix point of view, this is no condition, as the last $n - m$ columns of the matrix in Lemma 1.5.8 can be arbitrary. But there is no guarantee that we can find a T-invariant complement W_2, as we see from Corollary 1.5.9: The last $n - m$ columns of the matrix there must be of a special form. In fact, it is in general *not* true that every T-invariant subspace of V has a T-invariant complement. $\diamond$

Corollary 1.5.9 has a generalization to the situation of Definition 1.5.6.

Corollary 1.5.11. *Let $T : V \longrightarrow V$ be a linear transformation, and let $\{W_1, \ldots, W_k\}$ be a complementary set of subspaces of V, with W_i having dimension m_i, for each i. Let B_i be a basis of W_i, for each i, and let $B = B_1 \cup \ldots \cup B_k$ (so that B is a basis of V). Then each W_i is T-invariant if and only if*

$$Q = \begin{bmatrix} A_1 & & & & & \\ & A_2 & & & 0 & \\ & & A_3 & & & \\ & & & \ddots & & \\ & 0 & & & A_{k-1} & \\ & & & & & A_k \end{bmatrix}$$

is a block diagonal matrix, with A_i an m_i-by-m_i matrix, for each i.

Proof. We leave this as an exercise for the reader. $\square$

We have defined invariance under a linear transformation. We now define invariance under a matrix.

Definition 1.5.12. Let A be an n-by-n matrix. A subspace W of $\mathbb{C}^n$ is *invariant* under A, or *A-invariant*, if W is invariant under T_A, i.e., if $T_A(w) = Aw$ is in W for every vector w in W. $\diamond$

CHAPTER 2

The Structure of a Linear Transformation

Let V be a vector space and let $\mathcal{T} : V \longrightarrow V$ be a linear transformation. Our objective in this chapter is to prove that V has a basis $\mathcal{B}$ in which $[\mathcal{T}]_\mathcal{B}$ is a matrix in Jordan Canonical Form. Stated in terms of matrices, our objective is to prove that any square matrix A is similar to a matrix J in Jordan Canonical Form. (Throughout this chapter, we will feel free to switch between the language of matrices and the language of linear transformations. It is some times more convenient to use the one, and sometimes more convenient to use the other.)

Before we do this, we must introduce some basic structural notions: eigenvalues, eigenvectors, generalized eigenvectors, and the characteristic and minimum polynomials of a matrix or of a linear transformation.

2.1 EIGENVALUES, EIGENVECTORS, AND GENERALIZED EIGENVECTORS

In this section, we introduce eigenvalues, eigenvectors, and generalized eigenvectors, as well as the characteristic polynomial. First, we begin by considering matrices, and then we consider linear transformations.

Definition 2.1.1. Let A be an n-by-n matrix. If $v \neq 0$ is a vector such that, for some λ,

$$Av = \lambda v$$

then v is an *eigenvector* of A associated to the *eigenvalue* λ. ◇

Remark 2.1.2. We note that the definition of an eigenvalue/eigenvector can be expressed in an alternate form.

$$Av = \lambda v$$
$$Av = \lambda I v$$
$$(A - \lambda I)v = 0.$$

◇

Definition 2.1.3. For an eigenvalue a of A, we let $E(\lambda)$ denote the *eigenspace* of λ,

$$E(\lambda) = \{v \mid Av = \lambda v\} = \{v \mid (A - \lambda I)v = 0\} = \text{Ker}(A - \lambda I).$$

◇

We also note that the formulation in Remark 2.1.2 helps us find eigenvalues and eigenvectors. For if $(A - \lambda I)v = 0$ for a nonzero vector v, the matrix $A - \lambda I$ must be singular, and hence its determinant must be 0. This leads us to the following definition.

Definition 2.1.4. The *characteristic polynomial* of a matrix A is the polynomial $c_A(x) = \det(xI - A)$.

◇

Remark 2.1.5. This is the customary definition of the characteristic polynomial. But note that, if A is an n-by-n matrix, then the matrix $xI - A$ is obtained from the matrix $A - xI$ by multiplying each of its n rows by -1, and hence $\det(xI - A) = (-1)^n \det(A - xI)$. In practice, it is most convenient to work with $A - xI$ in finding eigenvectors, as this minimizes arithmetic. Furthermore, when we come to finding chains of generalized eigenvectors, it is (almost) essential to use $A - xI$, as using $xI - A$ would introduce lots of spurious minus signs.   ◇

Remark 2.1.6. Since A is an n-by-n matrix, we see from properties of the determinant that $c_A(x) = \det(xI - A)$ is a polynomial of degree n.   ◇

We have the following important and useful fact about characteristic polynomials.

Lemma 2.1.7. *Let A and B be similar matrices. Then they have the same characteristic polynomial, i.e.,* $c_A(x) = c_B(x)$.

Proof. Let $A = PBP^{-1}$. Then $xI - A = xI - PBP^{-1} = x(PIP^{-1}) - PBP^{-1} = P(xI)P^{-1} - PBP-1 = P(xI - B)P^{-1}$, so $c_A(x) = \det(xI - A) = \det(P(xI - B)P^{-1}) = \det(P)\det(xI - B)\det(P^{-1}) = \det(xI - B) = c_B(x)$. □

The practical use of Lemma 2.1.7 is that, if we wish to compute the characteristic polynomial $c_A(x)$ of the matrix A, we may instead "replace" A by a similar matrix B, whose characteristic polynomial $c_B(x)$ is easier to compute, and compute that. There is also an important theoretical use of this lemma, as we will see in Definition 2.1.23 below.

Here are some cases in which the characteristic polynomial is (relatively) easy to compute. These cases are important both in theory and in practice.

Lemma 2.1.8. *Let A be the diagonal matrix with diagonal entries $d_1, \ldots, d_n$. Then $c_A(x) = (x - d_1) \cdots (x - d_n)$.*

More generally, let A be an upper triangular matrix with diagonal entries $a_{1,1}, \ldots, a_{n,n}$. Then $c_A(x) = (x - a_{1,1}) \cdots (x - a_{n,n})$.

Proof. If A is the diagonal matrix with diagonal entries $d_1, \ldots, d_n$, then $xI - A$ is the diagonal matrix with diagonal entries $x - d_1, \ldots, x - d_n$. But the determinant of a diagonal matrix is the product of its diagonal entries. More generally, if A is an upper triangular matrix with diagonal entries $a_{1,1}, \ldots, a_{n,n}$, then $xI - A$ is an upper triangular matrix with diagonal entries $x - a_{1,1}, \ldots, x - a_{n,n}$. But the determinant of an upper triangular matrix is also the product of its diagonal entries. □

Lemma 2.1.9. *Let A be a block diagonal matrix,*

$$
A = \begin{bmatrix} A_1 & & & & & \\ & A_2 & & & 0 & \\ & & A_3 & & & \\ & & & \ddots & & \\ & 0 & & & A_{m-1} & \\ & & & & & A_m \end{bmatrix}.
$$

Then $c_A(x)$ is the product $c_A(x) = c_{A_1}(x) \cdots c_{A_m}(x)$. More generally, let A be a block upper triangular matrix

$$
A = \begin{bmatrix} A_1 & & & & & \\ & A_2 & & * & & \\ & & A_3 & & & \\ & & & \ddots & & \\ & 0 & & & A_{m-1} & \\ & & & & & A_m \end{bmatrix}.
$$

Then $c_A(x)$ is the product $c_A(x) = c_{A_1}(x) \cdots c_{A_m}(x)$.

Proof. This follows directly from properties of determinants. □

We now introduce two important quantities associated to an eigenvalue of a matrix A.

Definition 2.1.10. Let λ be an eigenvalue of a matrix A. The *algebraic multiplicity* of the eigenvalue λ is alg-mult$(\lambda) = $ the multiplicity of λ as a root of the characteristic polynomial $\det(\lambda I - A)$.

The *geometric multiplicity* of the eigenvalue λ is geom-mult$(\lambda) = $ the dimension of the eigenspace $E(\lambda)$.

It is common practice to use the word *multiplicity* (without a qualifier) to mean algebraic multiplicity.   ◇

There is an important relationship between these two quantities.

Lemma 2.1.11. *Let λ be an eigenvalue of a matrix A. Then*

$$1 \leq \text{geom-mult}(\lambda) \leq \text{alg-mult}(\lambda).$$

Proof. Let $W_1 = E(\lambda)$ be the eigenspace of A associated to the eigenvalue λ. By definition, W_1 has dimension $m = \text{geom-mult}(A)$.

Now the first inequality is immediate. Since λ is an eigenvalue of A, there is an associated eigenvector, so $W_1 \neq \{0\}$, and hence $m \geq 1$.

To prove the second inequality, we will have to do some work. We consider the linear transformation $\mathcal{T}_A$, where $\mathcal{T}_A(v) = Av$.

By definition, W_1 has dimension $m = $ geom-mult(λ). Then W_1 is a $\mathcal{T}_A$-invariant subspace of $\mathbb{C}^n$, as for any vector v in W_1, $\mathcal{T}_A(v) = Av = \lambda v$ is in W_1. Let W_2 be any complement of W_1 in $\mathbb{C}^n$. Then we can apply Lemma 1.5.8 to conclude that for an appropriate basis $\mathcal{B}$ of $\mathbb{C}^n$, B $= [\mathcal{T}]_\mathcal{B}$ is of the form

$$B = [\mathcal{T}]_\mathcal{B} = \begin{bmatrix} E & F \\ 0 & H \end{bmatrix},$$

where E is an m-by-m block.

But in this case, it is easy to see what E is. Since $W_1 = E(\lambda)$, by the definition of the eigenspace, $\mathcal{T}_A(v) = \lambda v$ for every v in W_1, so E is just the m-by-m scalar matrix λI. Then, applying Lemma 2.1.8, we see that $c_E(x) = (x - \lambda)^m$, and applying Lemma 2.1.9, we see that $c_B(x) = c_E(x)c_H(x) = (x - \lambda)^m c_H(x)$ so $c_B(x)$ is divisible by $(x - \lambda)^m$ (perhaps by a higher power of $x - \lambda$, and perhaps not, as we do not know anything about $c_H(x)$). But, by Lemma 2.1.7, $c_A(x) = c_B(x)$, so we see that alg-mult$(\lambda) \geq m = $ geom-mult(λ). □

Lemma 2.1.12. *Let λ be an eigenvalue of a matrix A with* alg-mult$(\lambda) = 1$. *Then* geom-mult$(\lambda) = 1$.

Proof. In this case, by Lemma 2.1.11, $1 \leq $ geom-mult$(\lambda) \leq $ alg-mult$(\lambda) = 1$, so geom-mult$(\lambda) = 1$. □,

In order to investigate the structure of A, it turns out to be important (indeed, crucial) to consider not only eigenvectors, but rather generalized eigenvectors, which we now define.

Definition 2.1.13. If $v \neq 0$ is a vector such that, for some λ,

$$(A - \lambda I)^j(v) = 0$$

for some positive integer j, then v is a *generalized eigenvector* of A associated to the eigenvalue λ. The smallest j with $(A - \lambda I)^j(v) = 0$ is the *index* of the generalized eigenvector v. ◇

Definition 2.1.14. For an eigenvalue λ of A, and a fixed integer j, we let $E_j(\lambda)$ be the set of vectors

$$E_j(\lambda) = \{v \mid (A - \lambda I)^j v = 0\} = \text{Ker}((A - \lambda I)^j).$$

◇

We note that $E_1(\lambda) = E(\lambda)$ is just the eigenspace of λ. We also note that $E_1(\lambda) \subseteq E_2(\lambda)$, as if $(A - \lambda I)v = 0$, then certainly $(A - \lambda I)^2 v = (A - \lambda I)((A - \lambda I)v) = (A - \lambda I)0 = 0$; similarly $E_2(\lambda) \subseteq E_3(\lambda)$, etc. It is convenient to set $E_0(\lambda) = \{0\}$, the vector space consisting of the 0 vector alone (and note that this is consistent with our notation, as $\text{Ker}((A - \lambda I)^0) = \text{Ker}(I) = \{0\}$).

Definition 2.1.15. For an eigenvalue λ of A, we let $E_\infty(\lambda)$ be the set of vectors

$$E_\infty(\lambda) = \{v \mid (A - \lambda I)^j v = 0 \text{ for some } j\} = \{v \mid v \text{ is in } E_j(\lambda) \text{ for some } j\}.$$

$E_\infty(\lambda)$ is called the *generalized eigenspace* of λ. ◇

Lemma 2.1.16. *Let λ be an eigenvector of A. For any j, $E_j(\lambda)$ is a subspace of $\mathbb{C}^n$. Also, $E_\infty(\lambda)$ is a subspace of $\mathbb{C}^n$.*

Proof. $E_j(\lambda)$ is the kernel of the linear transformation $\mathcal{T}_{(A-\lambda I)^j}$, and the kernel of a linear transformation is always a subspace. We leave the proof for $E_\infty(\lambda)$ to the reader. $\square$

In fact, these are not only subspaces but in fact A-invariant subspaces. Indeed, this is one of the most important ways in which invariant subspaces arise.

Lemma 2.1.17. *Let A be an n-by-n matrix and let λ be an eigenvalue of A. Then for any j, $E_j(\lambda)$ is an A-invariant subspace of $\mathbb{C}^n$. Also, $E_\infty(\lambda)$ is an A-invariant subspace of $\mathbb{C}^n$.*

Proof. Let v be any vector in $E_j(\lambda)$, and let $w = Av$. By definition, $(A - \lambda I)^j v = 0$. But then $(A - \lambda I)^j w = (A - \lambda I)^j (Av) = ((A - \lambda I)^j A)v = (A(A - \lambda I)^j)v = A((A - \lambda I)^j v) = A0 = 0$, so Aw is in $E_j(\lambda)$. Since any vector in $E_\infty(\lambda)$ is in $E_j(\lambda)$ for some j, this also shows that $E_\infty(\lambda)$ is A-invariant. $\square$

Here is a very useful computational lemma.

Lemma 2.1.18. *Let A be an n-by-n matrix and let λ be an eigenvalue of A. Let $f(x)$ be any polynomial. If v is any eigenvector of A associated to the eigenvalue λ, then $f(A)v = f(\lambda)v$. More generally, if v is any generalized eigenvector of A of index j associated to the eigenvalue λ, then $f(A)v = f(\lambda)v + v'$ where v' is in $E_{j-1}(\lambda)$.*

Proof. By definition, $Av = \lambda v$. Then $A^2 v = A(Av) = A(\lambda v) = \lambda(Av) = \lambda^2 v$, and similarly $A^i v = \lambda^i v$ for any positive integer i, from which the first claim follows.

Now for the second claim. By the division algorithm for polynomials, we can write $f(x)$ as $f(x) = (x - \lambda)q(x) + f(\lambda)$ for some polynomial $q(x)$ (i.e., $f(\lambda)$ is the remainder upon dividing $f(x)$ by $x - \lambda$). Then $f(A) = (A - \lambda I)q(A) + f(\lambda)I$. Now let v be a generalized eigenvector of A of index j associated to λ. Then $f(A)v = (A - \lambda I)q(A)v + f(\lambda)Iv = v' + f(\lambda)v$. Furthermore, $(A - \lambda I)^{j-1}v' = (A - \lambda I)^{j-1}((A - \lambda I)q(A))v = ((A - \lambda I)^{j-1}(A - \lambda I)q(A))v = ((A - \lambda I)^j q(A))v = (q(A)(A - \lambda I)^j)v = q(A)((A - \lambda I)^j v) = q(A)0 = 0$, so v' is in $E_{j-1}(\lambda)$. $\square$

From its definition, it appears that we may have to consider arbitrarily high powers of j in finding $E_\infty(\lambda)$. But in fact we do not, as we see from the next proposition.

Proposition 2.1.19. *Let A be an square matrix and let λ be an eigenvalue of A. Let $E_\infty(\lambda)$ be a subspace of $\mathbb{C}^n$ of dimension $d_\infty(\lambda)$. Then $E_\infty(\lambda) = E_j(\lambda)$ for some $j \leq d_\infty(\lambda)$. Furthermore, the smallest value of j for which $E_\infty(\lambda) = E_j(\lambda)$ is equal to the smallest value of j for which $E_{j+1}(\lambda) = E_j(\lambda)$.*

Proof. First note that since $E_\infty(\lambda)$ is a subspace of $\mathbb{C}^n$, $d_\infty(\lambda) \leq n$.

Let $d_i(\lambda) = \dim E_i(\lambda)$ for each positive integer i. Since $E_i(\lambda) \subseteq E_{i+1}(\lambda)$, we have that $d_i(\lambda) \leq d_{i+1}(\lambda)$. Furthermore, $d_1(\lambda) \geq 1$, as, by definition, $E_1(\lambda)$ is a nonzero vector space (there is some eigenvector in it), and $d_i(\lambda) \leq d_\infty(\lambda)$ for every i, as each $E_i(\lambda)$ is a subspace of $E_\infty(\lambda)$.

Thus we have a sequence of positive integers $1 \leq d_1(\lambda) \leq d_2(\lambda) < \ldots < d_\infty(\lambda)$, all of which are less than or equal to n, so this sequence cannot continually strictly increase, i.e., there is a value of j for which $d_j(\lambda) = d_{j+1}(\lambda)$. Furthermore, this must occur for $j \leq d_\infty(\lambda)$, as the longest possibly strictly increasing sequence of integers between 1 and $d_\infty(\lambda)$ is $1 < 2 < \ldots < d_\infty(\lambda)$, a sequence of length $d_\infty(\lambda)$.

We claim that $E_j(\lambda) = E_{j+1}(\lambda) = E_{j+2}(\lambda) = \ldots$, so $(d_j(\lambda) = d_{j+1}(\lambda) = d_{j+2}(\lambda) = \ldots)$ and $E_\infty(\lambda) = E_j(\lambda)$. To see this claim, first observe that $E_j(\lambda) \subseteq E_{j+1}(\lambda)$ and $\dim E_k(\lambda) = \dim E_{j+1}(\lambda)$, so $E_j(\lambda) = E_{j+1}(\lambda)$.

Now $E_{j+2}(\lambda) = \{v \mid (A - \lambda I)^{j+2} v = 0\} = \{v \mid (A - \lambda I)^{j+1}((A - \lambda I)v) = 0\}$. But the equality $E_j(\lambda) = E_{j+1}(\lambda)$ tells us that for any vector w, $(A - \lambda I)^{j+1} w = 0$ if and only if $(A - \lambda I)^j w = 0$; in particular, this is true for $w = (A - \lambda I)v$. Applying this, we see that $E_{j+2}(\lambda) = \{v \mid (A - \lambda I)^j((A - \lambda I)v) = 0\} = \{v \mid (A - \lambda I)^{j+1} v = 0\} = E_{j+1}(\lambda)$, etc. $\qquad \square$

With this proposition in hand, we can introduce a third important quantity associated to an eigenvalue a of A.

Definition 2.1.20. Let λ be an eigenvalue of a matrix A. Then max-ind(λ), the *maximum index of a generalized eigenvector of A associated to λ*, is the largest value of j such that A has a generalized eigenvector of index j associated to the eigenvalue λ. Equivalently, max-ind(λ) is the smallest value of j such that $E_j(\lambda) = E_\infty(\lambda)$. $\qquad \diamond$

Corollary 2.1.21. *Let A be an square matrix and let λ be an eigenvalue of A. Then* max-ind(λ) *is less than or equal to the dimension of $E_\infty(\lambda)$.*

Proof. This is just a restatement of Proposition 2.1.19 using the language of Definition 2.1.20. $\qquad \square$

Remark 2.1.22. It appears that we have a fourth important quantity associated to an eigenvalue λ of A, namely $d_\infty(\lambda)$. But we shall see below that $d_\infty(\lambda) = $ alg-mult(λ). We shall also see that the individual numbers $d_1(\lambda), d_2(\lambda), \ldots$ play an important role. $\qquad \diamond$

Now we wish to consider an arbitrary linear transformation $\mathcal{T} : V \longrightarrow V$ instead of just a square matrix A. We observe that almost everything we have done is exactly the same. For example, we can define the eigenspace $E(\lambda)$ of $\mathcal{T}$ by $E(\lambda) = \text{Ker}(\mathcal{T} - \lambda \mathcal{I})$, the spaces $E_j(\lambda)$ by $E_j(\lambda) = \text{Ker}((\mathcal{T} - \lambda \mathcal{I})^j)$, and similarly the generalized eigenspace $E_\infty(\lambda)$ by $E_\infty(\lambda) = \{v \mid v \text{ is in } E_j(\lambda) \text{ for some } j\}$. The one potential difficulty is in defining the characteristic polynomial, as it is not a priori clear what we mean by $\det(x\mathcal{I} - \mathcal{T})$. It is given as follows.

Definition 2.1.23. Let $\mathcal{T} : V \longrightarrow V$ be a linear transformation. Let $\mathcal{B}$ be a basis of V and set $A = [\mathcal{T}]_\mathcal{B}$. Then the *characteristic polynomial* of $\mathcal{T}$ is $c_\mathcal{T}(x) = c_A(x) = \det(xI - A)$. $\qquad \diamond$

Note that this definition makes sense. We have to choose a basis $\mathcal{B}$ to get the matrix A. If we choose another basis $\mathcal{C}$, we get another matrix B. But, by Theorem 1.2.8, A and B are similar, and then by Lemma 2.1.7, $c_A(x) = c_B(x)$. Thus it doesn't matter which basis we choose. In fact, we will be using this observation not only to know that the characteristic polynomial of $\mathcal{T}$ is well-defined, but also to be able to switch bases to find a basis in which it is easiest to compute. (Actually, we already have used it for this purpose in the proof of Lemma 2.1.11.)

Remark 2.1.24. Note that, following our previous notation, if $\mathcal{T} = \mathcal{T}_A$ is the linear transformation given by $\mathcal{T}_A(v) = Av$, then $c_{\mathcal{T}}(x) = c_{\mathcal{T}_A}(x) = c_A(x)$. ◇

2.2   THE MINIMUM POLYNOMIAL

In the previous section, we introduced an extremely important polynomial associated to a linear transformation or to a matrix: its characteristic polynomial. In this section, we introduce a second extremely important polynomial associated to a linear transformation or to a matrix: its minimum polynomial.

Proposition 2.2.1. *(1) Let A be an n-by-n matrix. Then there is a unique monic polynomial $m_A(x)$ of smallest degree with $m_A(A) = 0$.*

Proof. We begin by showing that there is some nonzero polynomial $f(x)$ with $f(A) = 0$.

Step 0: Note that the set $\{I\}$ consisting of the identity matrix alone is linearly independent.

Step 1: Consider the set of matrices $\{I, A\}$. If this set is linearly dependent, then we have a relation $c_0 I + c_1 A = 0$ with $c_1 \neq 0$, and so, setting $f(x) = c_0 + c_1 x$, then $f(A) = 0$.

Step 2: If $\{I, A\}$ is linearly independent, consider the set of matrices $\{I, A, A^2\}$. If this set is linearly dependent, then we have a relation $c_0 I + c_1 A + c_2 A^2 = 0$ with $c_2 \neq 0$, and so, setting $f(x) = c_0 + c_1 x + c_2 x^2$, then $f(A) = 0$.

Keep going. If this procedure stops at some step, then we have found such a polynomial $f(x)$. But this procedure must stop no later than step n^2, as at that step we have the set $\{I, A, \ldots, A^{n^2}\}$, a set of $n^2 + 1$ elements of the vector space of n-by-n matrices. But this vector space has dimension n^2, so this set must be linearly dependent.

We want a monic polynomial, but this is easy to arrange: If $f(x) = c_k x^k + c_{k-1} x^{k-1} = \ldots + c_0$, let $m_A(x) = (1/c_k) f(x) = x^k + a_{k-1} x^{k-1} + \ldots + a_0$ where $a_i = c_i / c_{k-1}$..

Finally, we want to show this polynomial is unique. So suppose we had another monic polynomial $g(x) = x^k + b_{k-1} x^{k-1} + \ldots + b_0$ with $g(A) = 0$. Let $h(x) = m_A(x) - g(x)$. Then $h(A) = m_A(A) - g(A) = 0 - 0 = 0$. But $h(x)$ has degree less than k (as the x^k terms cancel). However, k is the smallest degree of a nonzero polynomial $f(x)$ with $f(A) = 0$, so $h(x)$ must be the zero polynomial and hence $g(x) = m_A(x)$. □

This proposition leads us to a definition.

Definition 2.2.2. Let A be a matrix. The *minimum polynomial* of A is the unique monic polynomial $m_A(x)$ of smallest degree with $m_A(A) = 0$. ◇

Lemma 2.2.3. *Let A and B be similar matrices. Then $m_A(x) = m_B(x)$.*

Proof. Immediate from Lemma 1.4.5.   □

Remark 2.2.4. Let us emphasize the conclusion of Lemma 2.2.3. If we wish to find $m_A(x)$, we may replace A by a similar but hopefully simpler matrix B and instead find $m_B(x)$.   ◇

Corollary 2.2.5. *Let A be an n-by-n matrix. Then $m_A(x)$ is a polynomial of degree at most n^2.*

Proof. Immediate from the proof of Proposition 2.2.1.   □

Remark 2.2.6. We shall be investigating the minimum polynomial a lot more deeply, and we shall see that in fact the minimum polynomial of an n-by-n matrix has degree at most n.   ◇

The following proposition plays an important role.

Proposition 2.2.7. *Let A be an n-by-n matrix and let $f(x)$ be any polynomial with $f(A) = 0$. Then $f(x)$ is divisible by $m_A(x)$.*

Proof. Let $m_A(x)$ be a polynomial of degree k. By the division algorithm for polynomials, we may write $f(x) = m_A(x)q(x) + r(x)$ where either $r(x) = 0$ is the zero polynomial, or $r(x)$ is a nonzero polynomial of degree less than k. But then $r(x) = f(x) - m_A(x)q(x)$ so $r(A) = f(A) - m_A(A)q(A) = 0 - 0q(A) = 0 - 0 = 0$. If $r(x)$ were a nonzero polynomial, it would be a nonzero polynomial of degree less than k with $r(A) = 0$. By the definition of the minimum polynomial, no such polynomial can exist. Hence we must have $r(x) = 0$, so $f(x) = m_A(x)q(x)$ is a multiple of $m_A(x)$.   □

Now let us examine some examples of matrices, with a view towards determining their characteristic and minimum polynomials.

Example 2.2.8. (1) Let A be an n-by-n scalar matrix, $A = cI$. Then $A - cI = 0$, so if $f(x) = x - c$, then $f(A) = 0$. Clearly $f(x)$ has the smallest possible degree, as its degree is 1, so we see that $m_A(x) = x - c$, a polynomial of degree 1. On the other hand, by Lemma 2.1.8, $c_A(x) = (x - c)^n$.

(2) Let A be an n-by-n diagonal matrix with *distinct* diagonal entries d_1, ..., d_n. By Lemma 1.4.6, for any polynomial $f(x)$, $f(A)$ is the diagonal matrix with entries $f(d_1)$, $f(d_2)$, ..., $f(d_n)$. Thus, if $f_0(x) = (x - d_1) \cdots (x - d_n)$, then $f_0(A) = 0$. On the other hand, also by Lemma 1.4.6, if $f(x)$ is any polynomial with $f(A) = 0$, we must have $f(d_1) = \ldots = f(d_n) = 0$, so $f(x)$ must be divisible by the polynomial $f_0(x)$. Hence $f_0(x)$ must be the minimum polynomial of A, i.e., $m_A(x) = (x - d_1) \cdots (x - d_n)$. Again, by Lemma 2.1.8, $c_A(x) = (x - d_1) \cdots (x - d_n)$.

(3) Let A be an n-by-n diagonal matrix whose entries are not necessarily distinct. Suppose A has a diagonal entry d_1 that appears k_1 times, a diagonal entry d_2 that appears k_2 times, ..., and a diagonal entry d_m that appears k_m times. Then, exactly by the argument of (2), we have that $m_A(x) = (x - d_1) \cdots (x - d_m)$ and $c_A(x) = (x - d_1)^{k_1} \cdots (x - d_m)^{k_m}$.

(4) Let J be an n-by-n Jordan block with diagonal entry a. Let $\mathcal{E} = \{e_1, \ldots, e_n\}$ be the standard basis of $\mathbb{C}^n$. Then $(J - aI)e_1 = 0$ and $(J - aI)e_i = e_{i-1}$ for $i > 1$. But then also $(J -$

$aI)^2 e_i = 0$ for $i = 1, 2$ and $(J - aI)^2 e_i = 0$ for $i > 2$. Proceeding in this way, we find that $(J - aI)^{n-1} e_n = e_1$, so $(J - aI)^{n-1} \neq 0$, but that $(J - aI)^n e_i = 0$ for every i, which implies that $(J - aI)^n = 0$ (as $\mathcal{E}$ is a basis of $\mathbb{C}^n$). Thus we see that if $f_0(x) = (x - a)^n$, then $f_0(J) = 0$. Hence, by Proposition 2.2.6, $m_J(x)$ must divide $(x - a)^n$, so we must have $m_J(x) = (x - a)^k$ for some $k \leq n$. But we just observed that if $f(x) = (x - a)^{n-1}$, then $f(J) \neq 0$. Hence we must have $k = n$. Thus we see that $m_J(x) = (x - a)^n$. Also, $c_J(x) = (x - a)^n$. (Note that the first part of this argument is essentially the argument of the proof of Lemma 1.3.17.)

(6) Let A be an n-by-n upper triangular matrix with *distinct* diagonal entries $d_1, \ldots, d_n$. By Lemma 1.4.6, for any polynomial $f(x)$, $f(A)$ is an upper triangular matrix with entries $f(d_1)$, $f(d_2)$, ..., $f(d_n)$. Thus, if $f(x)$ is any polynomial with $f(A) = 0$, we must have $f(d_1) = \ldots = f(d_n) = 0$, so $f(x)$ must be divisible by the polynomial $f_0(x) = (x - d_1) \cdots (x - d_n)$. On the other hand, a very unenlightening computation shows that $f_0(x) = 0$. Hence $f_0(x)$ must be the minimum polynomial of A, i.e., $m_A(x) = (x - d_1) \cdots (x - d_n)$. Also, $c_A(x) = (x - d_1) \cdots (x - d_n)$.

(7) Let A be an n-by-n upper triangular matrix with diagonal entries d_1 appearing k_1 times, d_2 appearing k_2 times, ..., and d_m appearing k_m times. The argument of (6) shows that if $f(x)$ is any polynomial with $f(A) = 0$, then $f(x)$ must be divisible by $f_0(x) = (x - d_1) \cdots (x - d_m)$. On the other hand, if $f_1(x) = (x - d_1)^{k_1} \cdots (x - d_m)^{k_m}$, then a very unenlightening computation shows that $f_1(A) = 0$. Hence $f_0(x)$ divides $m_A(x)$, which divides $f_1(x)$. Thus we see that $m_A(x) = (x - d_1)^{j_1} \cdots (x - d_m)^{j_m}$ for some integers $1 \leq j_1 \leq k_1, \ldots, 1 \leq j_m \leq k_m$. Without any further information about A, we cannot specify the integers $\{j_i\}$ more precisely. But in any case, $c_A(x) = (x - d_1)^{k_1} \cdots (x - d_m)^{k_m}$. ◇

Remark 2.2.9. The easiest way to do the computations in Example 2.2.8 (6) and (7) is to use the following computational result, whose proof we leave to you: Let $M_1, \ldots, M_n$ be n n-by-n upper triangular matrices and suppose that, for each i between 1 and n, the ith diagonal entry of M_i is 0. Then $M_1 \cdots M_n = 0$. But no matter how you do it, this is a computation that does not yield much insight. We will see below *why* these results are true. ◇

For a polynomial $f(x) = (x - a_1)^{k_1} \cdots (x - a_m)^{k_m}$, we call the factors $x - a_i$ the *irreducible factors* of $f(x)$. We record here the two basic facts relating $m_A(x)$ and $c_A(x)$. For any square matrix A:

(1) $m_A(x)$ and $c_A(x)$ have the same irreducible factors; and

(2) $m_A(x)$ divides $c_A(x)$.

You can see that these are both true in all parts of Example 2.2.8. We will prove the first of these now, but we will have to do more work before we can prove the second.

Theorem 2.2.10. *Let A be any square matrix. Then its minimum polynomial $m_A(x)$ and its characteristic polynomial $c_A(x)$ have the same irreducible factors.*

Proof. First, we show that every irreducible factor of $c_A(x)$ is an irreducible factor of $m_A(x)$: Let $x - a$ be an irreducible factor of $c_A(x)$. Then a is an eigenvalue of A. Choose any associated eigenvector v. Since $m_A(A) = 0$, certainly $m_A(A)v = 0$. But by Lemma 2.1.18, $m_A(A)v = m_A(a)v$. Thus $0 = m_A(a)v$. Since $v \neq 0$, we must have $m_A(a) = 0$, and so $x - a$ divides $m_A(x)$.

Next, we show that every irreducible factor of $m_A(x)$ is an irreducible factor of $c_A(x)$: Let $x - a$ be an irreducible factor of $m_A(x)$. Write $m_A(x) = (x - a)q(x)$. Now $q(A) \neq 0$, as $q(x)$ is a polynomial of smaller degree that of $m_A(x)$. Thus $w = q(A)v \neq 0$ for some vector w. But then $(A - aI)w = (A - aI)(q(A)v) = ((A - aI)q(A))v = m_A(A)v = 0$. In other words, w is an eigenvector of A with associated eigenvalue a, and so $x - a$ divides $c_A(x)$.   □

Again, we may rephrase this in terms of linear transformations.

Definition 2.2.11. Let $T : V \longrightarrow V$ be a linear transformation. The *minimum polynomial* of T is the unique monic polynomial $m_T(x)$ of smallest degree with $m_T(T) = 0$.   ◇

Again, we should note that this definition makes sense. In this situation, let B be any basis of V and let $A = [T]_B$. Then A has a minimum polynomial, and $m_T(x) = m_A(x)$. Furthermore, this is well-defined, as if C is any other basis of V, and $B = [T]_C$, then A and B are similar, so $m_A(x) = m_B(x)$. Thus we see that $m_T(x)$ is independent of the choice of basis.

Remark 2.2.12. Note that, following our previous notation, if $T = T_A$ is the linear transformation given by $T_A(v) = Av$, then $m_T(x) = m_{T_A}(x) = m_A(x)$.   ◇

2.3   REDUCTION TO *BDBUTCD* FORM

Having begun by introducing some basic structural features of a linear transformation (or matrix), we now embark on our derivation of Jordan Canonical Form. This is a very long argument, and so we carry it out in steps. The first step, which we carry out in this section, is to show that we can get A to be in *BDBUTCD* form. Recall from Definition 1.3.8 that a matrix in *BDBUTCD* form is a block diagonal matrix whose blocks are upper triangular with constant diagonal.

We state this as a theorem.

Theorem 2.3.1. *Let V be a vector space and let $T : V \longrightarrow V$ be a linear transformation. Then V has a basis B in which $A = [T]_B$ is a matrix in BDBUTCD form.*

Proof. Let V have dimension n. We prove this by induction on n.

If $n = 1$, then the only possible linear transformation T is $T(v) = \lambda v$ for some λ, and then, in any basis B, $[T]_B$ is the 1-by-1 matrix $[\lambda]$, which is certainly in *BDBUTCD* form.

Now for the inductive step. In the course of this argument, we will sometimes be explicitly changing bases, and sometimes conjugating by nonsingular matrices. Of course, these have the same effect (cf. Remark 1.2.12), but sometimes one will be more convenient than the other.

Let $c_{\mathcal{T}}(x) = (x - \lambda_1)^{k_1} (x - \lambda_2)^{k_2} \cdots (x - \lambda_m)^{k_m}$. Let v_1 be an eigenvector of $\mathcal{T}$ with associated eigenvalue λ_1. Let W_1 be the subspace of V spanned by v_1. Since v_1 is an eigenvector of $\mathcal{T}$, W_1 is a $\mathcal{T}$-invariant subspace of V, by Lemma 2.1.16. Extend $\{v_1\}$ to a basis $\mathcal{D}$ of V. Then, by Lemma 1.5.8, we have that $[\mathcal{T}]_{\mathcal{D}}$ is a block upper triangular matrix,

$$Q = [\mathcal{T}]_{\mathcal{D}} = \begin{bmatrix} \lambda_1 & B \\ 0 & D \end{bmatrix},$$

where B is a 1-by-$(n-1)$ matrix and D is an $(n-1)$-by-$(n-1)$ matrix. By Lemma 2.1.9, $c_Q(x) = (x - \lambda_1) c_D(x)$, so $c_D(x) = (x - \lambda_1)^{k_1 - 1} (x - \lambda_2)^{k_2} \cdots (x - \lambda_m)^{k_m}$. We now apply the inductive hypothesis to conclude that there is an $(n-1)$-by-$(n-1)$ matrix P_0 such that $H = PDP^{-1}$ is in *BDBUTCD* form, and indeed with its first $k_1 - 1$ diagonal entries equal to λ_1. Let P be the n-by-n block diagonal matrix

$$P = \begin{bmatrix} 1 & 0 \\ 0 & P_0 \end{bmatrix}.$$

Then direct computation shows that

$$R = PQP^{-1} = \begin{bmatrix} \lambda_1 & F \\ 0 & H \end{bmatrix},$$

a block upper triangular matrix whose first k_1 diagonal entries are equal to λ_1. This is good, but not good enough. We need to "improve" R so that it is in *BDBUTCD* form.

Let $F = \begin{bmatrix} r_{1,2} \ldots r_{1,n} \end{bmatrix}$. The first $k_1 - 1$ entries of F are no problem. Since the first $k_1 - 1$ diagonal entries of H are equal to λ_1, we can "absorb" the upper left hand corner of R into a single k_1-by-k_1 block.

The remaining entries of F do present a problem. For simplicity of notation as we solve this problem, let us simply set $k = k_1$.

Now $R = [\mathcal{T}]_{\mathcal{C}}$ in some basis $\mathcal{C}$ of V. Let $\mathcal{C} = \{w_1, \ldots, w_n\}$. We will replace each w_t, for $t > k$, with a new vector u_s so as to eliminate the problematic entries of F. We do this one vector at a time, beginning with $t = k + 1$. To this end, let $u_t = w_t + c_t w_1$ where c_t is a constant yet to be determined. If R has entries $\{r_{i,j}\}$, then

$$\mathcal{T}(w_t) = r_{1,t} w_1 + r_{2,t} w_2 + \ldots r_{k,t} w_k + r_{t,t} w_t,$$

and

$$\mathcal{T}(c_t w_1) = c_t \lambda_1 w_1,$$

so

$$\begin{aligned} \mathcal{T}(u_t) = \mathcal{T}(w_t + c_t w_1) &= (r_{1,t} + c_t \lambda_1) w_1 + r_{2,t} w_2 + \ldots r_{k,t} w_k + r_{t,t} w_t \\ &= (r_{1,t} + c_t \lambda_1) w_1 + r_{2,t} w_2 + \ldots r_{k,t} w_k + r_{t,t} (u_t - c_t w_1) \\ &= (r_{1,t} + c_t \lambda_1 - c_t r_{t,t}) w_1 + r_{2,t} w_2 + \ldots r_{k,t} w_k + r_{t,t} u_t. \end{aligned}$$

The coefficient of w_1 in this expression is $r_{1,t} + c_t\lambda_1 - c_t r_{t,t} = r_{1,t} + c_t(\lambda_1 - r_{t,t})$. Note that $\lambda_1 - r_{t,t} \neq 0$, as the first $k = k_1$ diagonal entries of R are equal to λ_1, but the remaining diagonal entries of R are unequal to λ_1. Hence if we choose

$$c_t = -r_{1,t}/(\lambda_1 - r_{t,t})$$

we see that the w_1-coefficient of $T(u_t)$ is equal to 0. In other words, the matrix of T in the basis $\{w_1, \ldots, w_k, u_{k+1}, w_{k+2}, \ldots, w_n\}$ is of the same form as R, except that the entry in the $(1, k+1)$ position is 0. Now do the same thing successively for $t = k+2, \ldots, n$. When we have finished we obtain a basis $\mathcal{B} = \{w_1, \ldots, w_k, u_{k+1}, \ldots, u_n\}$ with $A = [T]_\mathcal{B}$ of the form

$$A = [T]_\mathcal{B} = \begin{bmatrix} M & 0 \\ 0 & N \end{bmatrix}$$

with M and N each in *BDBUTCD* form, in which case A itself is in *BDBUTCD* form, as required.
□

Actually, we proved a more precise result than we stated. Let us state that more precise result now.

Theorem 2.3.2. *Let V be a vector space and let $T : V \longrightarrow V$ be a linear transformation with characteristic polynomial $c_T(x) = (x - \lambda_1)^{k_1}(x - \lambda_2)^{k_2} \cdots (x - \lambda_m)^{k_m}$. Then V has a basis $\mathcal{B}$ in which $A = [T]_\mathcal{B}$ is a block diagonal matrix*

$$A = \begin{bmatrix} A_1 & & & & & \\ & A_2 & & & 0 & \\ & & A_3 & & & \\ & & & \ddots & & \\ & 0 & & & A_{m-1} & \\ & & & & & A_m \end{bmatrix}$$

where each A_i is a k_i-by-k_i upper triangular matrix with constant diagonal λ_i.

Proof. Clear from the proof of Theorem 2.3.1.
□

(Let us carefully examine the proof of Theorem 2.3.1. In the first part of the proof, we "absorbed" the entries in the first row into the first block, and we could do that precisely because the corresponding diagonal entries were all *equal* to λ_1. In the second part of the proof, we "eliminated" the entries in the first row, and we could do that precisely because the corresponding diagonal entries were all *unequal* to λ_1.)

Let us rephrase this theorem in terms of matrices.

Theorem 2.3.3. *Let A be an n-by-n matrix with characteristic polynomial $c_A(x) = (x - \lambda_1)^{k_1}(x - \lambda_2)^{k_2} \cdots (x - \lambda_m)^{k_m}$. Then A is similar to a block diagonal matrix*

$$B = \begin{bmatrix} B_1 & & & & & \\ & B_2 & & & 0 & \\ & & B_3 & & & \\ & & & \ddots & & \\ & 0 & & & B_{m-1} & \\ & & & & & B_m \end{bmatrix}$$

where each B_i is a k_i-by-k_i upper triangular matrix with constant diagonal λ_i.

   Proof. Immediate from Theorem 2.3.2. $\square$

   Although Theorem 2.3.2 is only an intermediate result on the way to Jordan Canonical Form, it already has some powerful consequences.

   One of these is the famous Cayley-Hamilton Theorem.

Theorem 2.3.4. *(Cayley-Hamilton) Let A be any square matrix. Then $c_A(A) = 0$.*

   Proof. If B is any matrix that is similar to A, then $c_A(x) = c_B(x)$ by Lemma 2.1.7, and also $c_A(A) = 0$ if and only if $c_B(B) = 0$, by Lemma 1.4.5. Thus, instead of showing that $c_A(A) = 0$, we need only show that $c_B(B) = 0$.

   Let A be similar to B, where B is as in the conclusion of Theorem 2.3.3,

$$B = \begin{bmatrix} B_1 & & & & & \\ & B_2 & & & 0 & \\ & & B_3 & & & \\ & & & \ddots & & \\ & 0 & & & B_{m-1} & \\ & & & & & B_m \end{bmatrix}$$

where each B_i is a k_i-by-k_i upper triangular matrix with constant diagonal λ_i.

   Then $c_B(B) = (B - \lambda_1 I)^{k_1}(B - \lambda_2 I)^{k_2} \cdots (B - \lambda_m I)^{k_m}$. Now $c_B(B)$ is a product of block diagonal matrices, and a product of block diagonal matrices is the block diagonal matrix whose blocks are the products of the individual blocks. However, the first factor $(B - \lambda_1 I)^{k_1}$ has its first block equal to $(B_1 - \lambda_1 I)^{k_1}$, which is equal to 0, by Lemma 1.3.16. Similarly, the second factor $(B - \lambda_2 I)^{k_2}$ has its second block equal to $(B_2 - \lambda_2 I)^{k_1}$, which is equal to 0, again by Lemma 1.3.16, and so forth. Hence the product $c_B(B)$ is a block diagonal matrix with all of its diagonal blocks equal to 0, i.e., $c_B(B) = 0$, as claimed. $\square$

   Examining the proof of the Cayley-Hamilton Theorem, we see we may derive the following result.

Corollary 2.3.5. *Let A be an n-by-n matrix with distinct eigenvalues $\lambda_1, \ldots, \lambda_m$. Let $E_\infty(\lambda_i)$ be the generalized eigenspace of λ_i, for each i. Then $\{E_\infty(\lambda_1), \ldots, E_\infty(\lambda_m)\}$ is a complementary set of A-invariant subspaces of $\mathbb{C}^n$. In this situation, $\mathbb{C}^n$ is the direct sum $V = E_\infty(\lambda_1) \oplus \ldots \oplus E_\infty(\lambda_m)$.*

Proof. Let $A = PBP^{-1}$ as in Theorem 2.3.3 and let $\mathcal{B}$ be the basis of $\mathbb{C}^n$ consisting of the columns of P. Write $\mathcal{B} = \mathcal{B}_1 \cup \ldots \cup \mathcal{B}_m$, where $\mathcal{B}_1$ consists of the first m_1 vectors in $\mathcal{B}$, $\mathcal{B}_2$ consists of the next m_2 vectors in $\mathcal{B}$, etc. Let W_i be the subspace of $\mathbb{C}^n$ spanned by $\mathcal{B}_i$. Certainly $\{W_1, \ldots, W_m\}$ is a complementary set of subspaces of $\mathbb{C}^n$, and so $\mathbb{C}^n = W_1 \oplus \ldots \oplus W_m$. We claim that in fact $W_i = E_\infty(\lambda_i)$, for each i, which completes the proof. (Note that each $E_\infty(\lambda_i)$ is A-invariant, by Lemma 2.1.17. But also, we can see directly that each W_i is A-invariant, by Lemma 1.5.8.)

To see this claim, let $\mathcal{T}_i$ be the restriction of $\mathcal{T} = \mathcal{T}_A$ to W_i. Then $B_i = [\mathcal{T}_i]_{\mathcal{B}_i}$. Now B_i is a k_i-by-k_i upper triangular matrix with constant diagonal λ_i, so, by Lemma 1.3.16, $(B_i - \lambda_i I)^{k_i} = 0$. Thus any vector w_i in W_i is in $E_{k_i}(\lambda_i) = E_\infty(\lambda_i)$, i.e., $W_i \subseteq E_\infty(\lambda_i)$. On the other hand, if v in V is not in W_i, then, writing $v = w_1 + \ldots + w_m$, with w_j in W_j for each j, there must be some value of $j \neq i$ with $w_j \neq 0$. But $(B_j - \lambda_i I)$ is a block upper triangular matrix with constant diagonal $\lambda_j - \lambda_i \neq 0$, so in particular $(B_j - \lambda_i I)$ is nonsingular, as is any of its powers. Hence $(\mathcal{T}_j - \lambda_i \mathcal{I})^k w_j \neq 0$ no matter what k is. But then also $(\mathcal{T} - \lambda_i \mathcal{I})^k v \neq 0$. Thus $E_\infty(\lambda_i) \subseteq W_i$, and hence these two subspaces are the same. $\qquad\square$

The next consequence of Theorem 2.3.2 is the relationship between the minimum polynomial $m_A(x)$ and the characteristic polynomial $c_A(x)$.

Theorem 2.3.6. *For any square matrix A:*

(1) $m_A(x)$ and $c_A(x)$ have the same irreducible factors; and

(2) $m_A(x)$ divides $c_A(x)$.

Proof. We proved (1) in Theorem 2.2.10. By the Cayley-Hamilton Theorem, $c_A(A) = 0$, so (2) follows immediately from Proposition 2.2.7. $\qquad\square$

Corollary 2.3.7. *Let A be an n-by-n matrix. Then $m_A(x)$ is a polynomial of degree at most n.*

Proof. By Theorem 2.3.6, $m_A(x)$ divides $c_A(x)$, and $c_A(x)$ is a polynomial of degree n. $\qquad\square$

Let us now restate Theorem 2.3.6 in a more handy form.

Corollary 2.3.8. *Let A be a square matrix with characteristic polynomial $c_A(x) = (x - \lambda_1)^{k_1}(x - \lambda_2)^{k_2} \cdots (x - \lambda_m)^{k_m}$. Then A has minimum polynomial $m_A(x) = (x - \lambda_1)^{j_1}(x - \lambda_2)^{j_2} \cdots (x - \lambda_m)^{j_m}$ for some integers $j_1, j_2, \ldots, j_m$ with $1 \leq j_i \leq k_i$ for each i.*

Proof. Immediate from Theorem 2.3.6. $\qquad\square$

This corollary has two special cases that are worth stating separately.

Corollary 2.3.9. *Let A be a square matrix with distinct eigenvalues $\lambda_1, \ldots, \lambda_n$. Then $m_A(x) = c_A(x) = (x - \lambda_1)(x - \lambda_2) \cdots (x - \lambda_n)$.*

Proof. Immediate from Corollary 2.3.8. □

Corollary 2.3.10. *Let A be an n-by-n square matrix. The following are equivalent:*

(1) $c_A(x) = (x - \lambda)^n$ for some λ; and

(2) $m_A(x) = (x - \lambda)^k$ for some λ and for some $k \leq n$.

Proof. Immediate from Corollary 2.3.8. □

Here are two final consequences of Theorem 2.3.2.

Corollary 2.3.11. *Let A be a square matrix and let λ be an eigenvalue of A. Then the dimension of the generalized eigenspace of λ is equal to the algebraic multiplicity* alg-mult(λ).

Proof. Examining the proof of Corollary 2.3.5, we see that $W_i = E_\infty(\lambda_i)$ has dimension $m_i = $ alg-mult(λ_i), for any i. □

Corollary 2.3.12. *Let A be a square matrix with minimum polynomial $m_A(x) = (x - \lambda_1)^{j_1}(x - \lambda_2)^{j_2} \cdots (x - \lambda_m)^{j_m}$. Then $j_i = $ max-ind(λ_i) for each i,*

Proof. Examining the proof of the Cayley-Hamilton Theorem, we see that j_i is the smallest exponent such that the block $(B_i - \lambda_i I)^{j_i} = 0$. But, in the notation of the proof of Corollary 2.3.5, W_i is the generalized eigenspace of A associated to the eigenvalue λ_i, so j_i is the smallest exponent such that $(A - \lambda_i I)^{j_i} w = 0$ for every generalized eigenvector w associated to the eigenvalue λ_i, and by definition that is max-ind(λ_i). □

2.4 THE DIAGONALIZABLE CASE

We pause in our general development of Jordan Canonical Form to consider the important special case of diagonalizable matrices, or diagonalizable linear transformations.

Definition 2.4.1. Let A be an n-by-n square matrix. Then A is *diagonalizable* if A is similar to a diagonal matrix B. ◇

Theorem 2.4.2. *Let A be a square matrix with characteristic polynomial $c_A(x) = (x - \lambda_1)^{k_1}(x - \lambda_2)^{k_2} \cdots (x - \lambda_m)^{k_m}$. The following are equivalent:*

(1) A is diagonalizable.

(2a) For each eigenvalue λ_i of A, geom-mult(λ_i) = alg-mult(λ_i).

(2b) For each eigenvalue λ_i of A, $E_\infty(\lambda_i) = E_1(\lambda_i)$.

(2c) For each eigenvalue λ_i of A, max-ind(λ_i) = 1.

(2d) geom-mult$(\lambda_1) + \ldots +$ geom-mult$(\lambda_k) = n$.

(3) The minimum polynomial $m_A(x)$ of A is a product of distinct linear factors, $c_A(x) = (x - \lambda_1)(x - \lambda_2) \cdots (x - \lambda_m)$.

Proof. First let us see that conditions (2a), (2b), (2c), and (2d) are all equivalent.

We know that $E_1(\lambda_i)$ is a subspace of $E_\infty(\lambda_i)$, so these two subspaces of $\mathbb{C}^n$ are equal if and only if they have the same dimension. But by definition $E_1(\lambda_i)$ has dimension geom-mult(λ_i), and by Corollary 2.3.11 $E_\infty(\lambda_i)$ has dimension alg-mult(λ_i). Thus (2a) and (2b) are equivalent. But $E_\infty(\lambda_i) = E_1(\lambda_i)$ if and only if every generalized eigenvector of A associated to the eigenvalue λ_i is in fact an eigenvector of A, i.e., if and only if max-ind$(\lambda_i) = 1$, so (2b) and (2c) are equivalent. Also, $1 \leq$ geom-mult$(\lambda_i) \leq$ alg-mult(λ_i) for each i, and we know that alg-mult$(\lambda_1) + \ldots +$ alg-mult$(\lambda_k) = n$, so geom-mult$(\lambda_1) + \ldots +$ geom-mult$(\lambda_k) = n$ if and only if geom-mult$(\lambda_i) =$ alg-mult(λ_i) for each i, so (2d) and (2a) are equivalent.

Furthermore, (2c) and (3) are equivalent by Corollary 2.3.12.

Now we must relate all these equivalent conditions to diagonalizability.

Suppose that A is diagonalizable, $A = PBP^{-1}$. Since $c_B(x) = c_A(x)$, B must be a diagonal matrix with k_i entries of λ_i, for each i. We may assume (after conjugating by a further matrix, if necessary), that the entries appear in that order, so we may regard B as being a *BDBSM* matrix (i.e., a block diagonal matrix with blocks scalar matrices), where the ith block B_i is the k_i-by-k_i scalar matrix $B_i = \lambda_i I$. If $\mathcal{B}$ is the basis of $\mathbb{C}^n$ consisting of the columns of P, and W_1 is the subspace of $\mathbb{C}^n$ spanned by the first k_1 columns of P, W_2 is the subspace of $\mathbb{C}^n$ spanned by the next k_2 columns of P, etc., then $W_i = E_\infty(\lambda_i) = E_1(\lambda_i)$ for each i.

On the other hand, suppose that $E_\infty(\lambda_i) = E_1(\lambda_i)$ for each i. Then $V = E_\infty(\lambda_1) \oplus \ldots \oplus E_\infty(\lambda_m) = E_1(\lambda_1) \oplus \ldots \oplus E_1(\lambda_m)$. Let P be the matrix whose first k_1 columns are a basis for $E_1(\lambda_1)$, whose next k_2 columns are a basis for $E_1(\lambda_1)$, etc. Since $Av = \lambda_i v$ for every vector v in $E_1(\lambda_i)$, we see that $A = PBP^{-1}$ where B is the *BDBSM* matrix with diagonal blocks $B_1, \ldots, B_m$, where B_i is the k_i-by-k_i scalar matrix $B_i = \lambda_i I$. In particular, B is a diagonal matrix. $\square$

There is an important special case when diagonalizability is automatic.

Corollary 2.4.3. *Let A be a square matrix with distinct eigenvalues. Then A is diagonalizable.*

Proof. Immediate from Theorem 2.4.2 and Corollary 2.3.9. $\square$

Translated from matrices into linear transformations, we obtain the following results:

Definition 2.4.4. Let $\mathcal{T} : V \longrightarrow V$ be a linear transformation. Then $\mathcal{T}$ is *diagonalizable* if V has a basis $\mathcal{B}$ with $[\mathcal{T}]_{\mathcal{B}}$ a diagonal matrix.   ◇

Theorem 2.4.5. *Let* $\mathcal{T} : V \longrightarrow V$ *be a linear transformation with characteristic polynomial* $c_{\mathcal{T}}(x) = (x - \lambda_1)^{k_1}(x - \lambda_2)^{k_2} \cdots (x - \lambda_m)^{k_m}$. *The following are equivalent:*

(1) $\mathcal{T}$ *is diagonalizable.*

(2a) For each eigenvalue λ_i *of* $\mathcal{T}$, geom-mult$(\lambda_i) = $ alg-mult(λ_i).

(2b) For each eigenvalue λ_i *of* $\mathcal{T}$, $E_\infty(\lambda_i) = E_1(\lambda_i)$.

(2c) For each eigenvalue λ_i *of* $\mathcal{T}$, max-ind$(\lambda_i) = 1$.

(2d) geom-mult$(\lambda_1) + \ldots + $ geom-mult$(\lambda_k) = n$.

(3) The minimum polynomial $m_{\mathcal{T}}(x)$ *of* $\mathcal{T}$ *is a product of distinct linear factors,* $c_{\mathcal{T}}(x) = (x - \lambda_1)(x - \lambda_2) \cdots (x - \lambda_m)$.

Proof. Immediate from Theorem 2.4.2.   □

Corollary 2.4.6. *Let* $\mathcal{T} : V \longrightarrow V$ *be a linear transformation with distinct eigenvalues. Then* $\mathcal{T}$ *is diagonalizable.*

Proof. Immediate from Corollary 2.4.3.   □

2.5   REDUCTION TO JORDAN CANONICAL FORM

In this section, we complete our objective of showing that every linear transformation has a Jordan Canonical Form. In light of Corollary 2.3.5, we will work one eigenvalue at a time.

Let us fix a linear transformation $\mathcal{T} : V \longrightarrow V$. Consider a generalized eigenvector v_k of index k associated to an eigenvalue λ of $\mathcal{T}$, and set

$$v_{k-1} = (\mathcal{T} - \lambda \mathcal{I})v_k.$$

We claim that v_{k-1} is a generalized eigenvector of index $k - 1$ associated to the eigenvalue λ of $\mathcal{T}$. To see this, note that

$$(\mathcal{T} - \lambda \mathcal{I})^{k-1}v_{k-1} = (\mathcal{T} - \lambda \mathcal{I})^{k-1}(\mathcal{T} - \lambda \mathcal{I})v_k = (\mathcal{T} - \lambda \mathcal{I})^k v_k = 0$$

but

$$(\mathcal{T} - \lambda \mathcal{I})^{k-2}v_{k-1} = (\mathcal{T} - \lambda \mathcal{I})^{k-2}(\mathcal{T} - \lambda \mathcal{I})v_k = (\mathcal{T} - \lambda \mathcal{I})^{k-1}v_k \neq 0.$$

Proceeding in this way, we may set

$$v_{k-2} = (\mathcal{T} - \lambda \mathcal{I})v_{k-1} = (\mathcal{T} - \lambda \mathcal{I})^2 v_k$$
$$v_{k-3} = (\mathcal{T} - \lambda \mathcal{I})v_{k-2} = (\mathcal{T} - \lambda \mathcal{I})^2 v_{k-1} = (\mathcal{T} - \lambda \mathcal{I})^3 v_k$$
$$\vdots$$
$$v_1 = (\mathcal{T} - \lambda \mathcal{I})v_2 = \cdots = (\mathcal{T} - \lambda \mathcal{I})^{k-1} v_k$$

and note that each v_i is a generalized eigenvector of index i associated to the eigenvalue λ of $\mathcal{T}$. A collection of generalized eigenvectors obtained in this way gets a special name.

Definition 2.5.1. If $\{v_1, \ldots, v_k\}$ is a set of generalized eigenvectors associated to the eigenvalue λ of $\mathcal{T}$, such that v_k is a generalized eigenvector of index k, and also

$$v_{k-1} = (\mathcal{T} - \lambda \mathcal{I})v_k, \quad v_{k-2} = (\mathcal{T} - \lambda \mathcal{I})v_{k-1}, \quad v_{k-3} = (\mathcal{T} - \lambda \mathcal{I})v_{k-2},$$
$$\cdots, \quad v_2 = (\mathcal{T} - \lambda \mathcal{I})v_3, \quad v_1 = (\mathcal{T} - \lambda \mathcal{I})v_2,$$

then $\{v_1, \ldots, v_k\}$ is called a *chain* of generalized eigenvectors of length k associated to the eigenvalue λ of $\mathcal{T}$. The vector v_k is called the *top* of the chain and the vector v_1 (which is an ordinary eigenvector) is called the *bottom* of the chain.   ◇

Remark 2.5.2. It is important to note that a chain of generalized eigenvectors $\{v_1, \ldots, v_k\}$ is entirely determined by the vector v_k at the top of the chain. For once we have chosen v_k, there are no other choices to be made: the vector v_{k-1} is determined by the equation $v_{k-1} = (\mathcal{T} - \lambda \mathcal{I})v_k$; then the vector v_{k-2} is determined by the equation $v_{k-2} = (\mathcal{T} - \lambda \mathcal{I})v_{k-1}$; etc.   ◇

Lemma 2.5.3. *Let $\{v_1, \ldots, v_k\}$ be a chain of generalized eigenvectors of length k associated to the eigenvalue λ of the linear transformation $\mathcal{T}$. Then $\{v_1, \ldots, v_k\}$ is linearly independent.*

Proof. For simplicity, let us set $\mathcal{S} = \mathcal{T} - \lambda \mathcal{I}$.
Suppose we have a linear combination

$$c_1 v_1 + c_2 v_2 + \cdots + c_{k-1} v_{k-1} + c_k v_k = 0.$$

We must show each $c_i = 0$.

By the definition of a chain, $v_{k-i} = \mathcal{S}^i v_k$ for each i, so we may write this equation as

$$c_1 \mathcal{S}^{k-1} v_k + c_2 \mathcal{S}^{k-2} v_k + \cdots + c_{k-1} \mathcal{S} v_k + c_k v_k = 0.$$

Now let us multiply this equation on the left by $\mathcal{S}^{k-1}$. Then we obtain the equation

$$c_1 \mathcal{S}^{2k-2} v_k + c_2 \mathcal{S}^{2k-3} v_k + \cdots + c_{k-1} \mathcal{S}^k v_k + c_k \mathcal{S}^{k-1} v_k = 0.$$

Now $\mathcal{S}^{k-1} v_k = v_1 \neq 0$. However, $\mathcal{S}^k v_k = 0$, and then also $\mathcal{S}^{k+1} v_k = \mathcal{S} \mathcal{S}^k v_k = \mathcal{S}(0) = 0$, and then similarly $\mathcal{S}^{k+2} v_k = 0, \ldots, \mathcal{S}^{2k-2} v_k = 0$, so every term except the last one is zero and this equation becomes

$$c_k v_1 = 0.$$

Since $v_1 \neq 0$, this shows $c_k = 0$, so our linear combination is

$$c_1 v_1 + c_2 v_2 + \cdots + c_{k-1} v_{k-1} = 0.$$

Repeat the same argument, this time multiplying by $\mathcal{S}^{k-2}$ instead of $\mathcal{S}^{k-1}$. Then we obtain the equation

$$c_{k-1} v_1 = 0.$$

and, since $v_1 \neq 0$, this shows that $c_{k-1} = 0$ as well. Keep going to get

$$c_1 = c_2 = \cdots = c_{k-1} = c_k = 0,$$

so $\{v_1, \ldots, v_k\}$ is linearly independent. □

Proposition 2.5.4. *Let $\mathcal{T} : V \longrightarrow V$ be a linear transformation with $c_{\mathcal{T}}(x) = (x - \lambda)^k$. Suppose that* max-ind$(\mathcal{T}) = k$ *and let v_k be any generalized eigenvector of index k associated to the eigenvalue λ of $\mathcal{T}$. Let $\mathcal{B} = \{v_1, \ldots, v_k\}$, a chain of generalized eigenvectors of length k. Then $\mathcal{B}$ is a basis of V and $J = [\mathcal{T}]_{\mathcal{B}}$ is given by*

$$J = \begin{bmatrix} \lambda & 1 & & & & \\ & \lambda & 1 & & & \\ & & \lambda & 1 & & \\ & & & \ddots & \ddots & \\ & & & & \lambda & 1 \\ & & & & & \lambda \end{bmatrix},$$

i.e., J is a single k-by-k Jordan block with diagonal entries equal to λ.

Proof. By Lemma 2.5.3, $\mathcal{B}$ is a linearly independent set of k vectors in the vector space V of dimension k, so is a basis of V. Now the ith column of $[\mathcal{T}]_{\mathcal{B}}$ is $[\mathcal{T}(v_i)]_{\mathcal{B}}$. By the definition of a chain,

$$\begin{aligned} \mathcal{T}(v_i) &= (\mathcal{T} - \lambda\mathcal{I} + \lambda\mathcal{I})v_i \\ &= (\mathcal{T} - \lambda\mathcal{I})v_i + \lambda\mathcal{I}v_i \\ &= v_{i-1} + \lambda v_i \text{ for } i > 1, = \lambda v_i \text{ for } i = 1, \end{aligned}$$

so for $i > 1$

$$[\mathcal{T}(v_i)]_{\mathcal{B}} = \begin{bmatrix} 0 \\ \vdots \\ 1 \\ \lambda \\ 0 \\ \vdots \end{bmatrix}$$

with 1 in the $(i-1)^{st}$ position, λ in the ith position, and 0 elsewhere, and $[\mathcal{T}(v_1)]_{\mathcal{B}}$ is similar, except that λ is in the 1st position, (there is no entry of 1), and every other entry is 0. Assembling these vectors, and recalling Definition 1.3.11, we see that the matrix $J = [\mathcal{T}]_\mathcal{B}$ has the form of a single k-by-k Jordan block with diagonal entries equal to λ. $\qquad\square$

Here is the goal to which we have been heading. Before stating this theorem, we recall from Definition 1.3.12 that a matrix in Jordan Canonical Form is a block diagonal matrix, with each block a Jordan block.

Theorem 2.5.5. *Let V be a vector space and let $\mathcal{T} : V \longrightarrow V$ be a linear transformation. Then V has a basis $\mathcal{B}$ in which $A = [\mathcal{T}]_\mathcal{B}$ is a matrix in Jordan Canonical Form.*

Proof. Let V have dimension n. We prove this by complete induction on n.

If $n = 1$, then the only possible linear transformation $\mathcal{T}$ is $\mathcal{T}(v) = \lambda v$ for some λ, and then, in any basis $\mathcal{B}$, $[\mathcal{T}]_\mathcal{B}$ is the 1-by-1 matrix $[\lambda]$, which is certainly in Jordan Canonical Form.

Now for the inductive step. In the course of this argument, we will sometimes be explicitly changing bases, and sometimes conjugating by nonsingular matrices. Of course, these have the same effect (cf. Remark 1.2.12), but sometimes one will be more convenient than the other.

As a first step, we know by Theorem 2.3.1 that V has a basis $\mathcal{C}$ with $[\mathcal{T}]_\mathcal{C}$ a matrix in *BDBUTCD* form. More precisely, let $c_\mathcal{T}(x) = (x - \lambda_1)^{k_1}(x - \lambda_2)^{k_2} \cdots (x - \lambda_m)^{k_m}$. Let $W_i = E_\infty(\lambda_i)$, for each i. We may write $\mathcal{C} = \mathcal{C}_1 \cup \ldots \cup \mathcal{C}_m$ where $\mathcal{C}_i$ is a basis of W_i, for each i. Let $\mathcal{T}_i$ denote the restriction of $\mathcal{T}$ to W_i.

First, consider W_1 and $\mathcal{T}_1$. Let $k = k_1$. Let $j = j_1 = \max\text{-ind}(\lambda_1)$. Choose a generalized eigenvector w_j of index j in W_1 and form a chain of generalized eigenvectors $\mathcal{D}_{1,1} = \{w_1, \ldots, w_j\}$. This set is a basis for a subspace $W_{1,1}$ of W_1. If $j_1 = k_1$ then $W_{1,1} = W_1$ and $\mathcal{D}_{1,1}$ is a basis for W_1. In that case, we set $\mathcal{B}_1 = \mathcal{D}_{1,1}$, and we are finished with the eigenvalue λ_1. Otherwise, let $W_{1,2}$ be a complement of $W_{1,1}$ in W_1, and let $\mathcal{D}_{1,2}$ be a basis for $W_{1,2}$. Let $\mathcal{D} = \mathcal{D}_{1,1} \cup \mathcal{D}_{1,2}$. Finally, let $\mathcal{T}_{1,1}$ be the restriction of $\mathcal{T}_1$ to $W_{1,1}$. By Proposition 2.5.4, we know that $[\mathcal{T}_{1,1}]_{\mathcal{D}_{1,1}} = J_1$ is a j-by-j Jordan block with diagonal entries of λ_1. Furthermore, by Lemma 1.5.8, we also know that $[\mathcal{T}]_\mathcal{D}$ is a block upper triangular matrix,

$$Q = [\mathcal{T}]_\mathcal{D} = \begin{bmatrix} J_1 & B \\ 0 & D \end{bmatrix},$$

where B is a j-by-$(k - j)$ matrix and D is an $(k - j)$-by-$(k - j)$ matrix. We now apply the inductive hypothesis to conclude that there is a $(k - j)$-by-$(k - j)$ matrix P_0 such that $H = P D P^{-1}$ is in Jordan Canonical Form (with all diagonal entries equal to λ_1). Let P be the n-by-n block diagonal matrix

$$P = \begin{bmatrix} I & 0 \\ 0 & P_0 \end{bmatrix}.$$

Then direct computation shows that

$$R = PQP^{-1} = \begin{bmatrix} J_1 & F \\ 0 & H \end{bmatrix},$$

a block upper triangular matrix all of whose diagonal blocks are Jordan blocks (with all diagonal entries equal to λ_1). Denote these Jordan blocks by J_1, J_2, ... This is good, but not good enough. We need to "improve" R so that it is in Jordan Canonical Form. In order to do this, we need to find a new basis of W_1 that has the effect of eliminating the nonzero entries of F.

Now $R = [\mathcal{T}_1]_{\mathcal{C}_1}$ in some basis $\mathcal{C}_1$ of W_1. Let $\mathcal{C}_1 = \{w_1, \ldots, w_n\}$. We will replace each w_s, for $s > j$, with a new vector u_s so as to eliminate the problematic entries of F. We do this one vector at a time, beginning with $s = j + 1$, but do our replacement so as to simultaneously remove all of the nonzero entries in that column of F.

Let the entries in the first column of F, or equivalently, in the sth column of R, be $r_{1,s}$, ..., $r_{j,s}$. We begin with the key observation that $r_{j,s} = 0$. To see this, note that, by the definition of $R = [\mathcal{T}_1]_{\mathcal{C}_1}$, we have that

$$\mathcal{T}_1(w_s) = r_{1,s}w_1 + r_{2,s}w_2 + \ldots + r_{j,s}w_j + r_{s,s}w_s$$
$$= r_{1,s}w_1 + r_{2,s}w_2 + \ldots + r_{j,s}w_j + \lambda_1 w_s$$

so, setting $\mathcal{S} = \mathcal{T}_1 - \lambda_1 \mathcal{I}$ for simplicity,

$$\mathcal{S}(w_s) = r_{1,s}w_1 + r_{2,s}w_2 + \ldots + r_{j,s}w_j.$$

But by the definition of a chain of generalized eigenvectors,

$$\mathcal{S}^{i_2}(w_{i_1}) = 0$$

for $1 \leq i_1 \leq j$ and any $i_2 \geq i_1$. In particular, let $i_2 = j - 1$. We then have

$$\mathcal{S}^j(w_i) = \mathcal{S}^{j-1}\mathcal{S}(w_s) = \mathcal{S}^{j-1}(r_{1,s}w_1 + r_{2,s}w_2 + \ldots + r_{j,s}w_j)$$
$$= 0 + 0 + \ldots + r_{j,s}\mathcal{S}^{j-1}(w_j) = r_{j,s}w_1.$$

However, we chose $j = \text{max-ind}(\lambda_1)$ to be the highest index of any generalized eigenvector associated to the eigenvalue λ_1. Thus $\mathcal{S}^j(w) = 0$ for every w in W_1, and in particular $\mathcal{S}^j(w_s) = 0$, so we have that $r_{j,s} = 0$.

With this observation in hand, it is easy to find the desired vector u_s. We simply set

$$u_s = w_s - (r_{1,s}w_2 + \ldots + r_{j-1,s}w_j),$$

and compute that

$$\mathcal{S}(u_s) = \mathcal{S}(w_s - (r_{1,s}w_2 + \ldots + r_{j-1,s}w_j))$$
$$= \mathcal{S}(w_s) - \mathcal{S}(r_{1,s}w_2 + \ldots + r_{j-1,s}w_j))$$
$$= (r_{1,s}w_1 + \ldots + r_{j-1,s}w_{j-1})) - (r_{1,s}w_1 + \ldots + r_{j-1,s}w_{j-1})) = 0,$$

so

$$\mathcal{T}(u_s) = (\mathcal{S} + \lambda_1 \mathcal{I})(u_s) = \lambda_1 u_s.$$

Thus in the basis $C_1' = \{w_1, \ldots, w_j, u_{j+1}, w_{j+2}, \ldots, w_k\}$ of W_1, we have $[T_1]_{C_1'} = R'$ is a matrix

$$R' = \begin{bmatrix} J_1 & F' \\ 0 & H \end{bmatrix},$$

where J_1 and H are as before, but now every entry in the first column of F' is 0. Continue in this fashion to obtain a basis B_1 of W_1 with $[T_1]_{B_1}$ of the form

$$\begin{bmatrix} J_1 & 0 \\ 0 & H \end{bmatrix},$$

a matrix in Jordan Canonical Form with all diagonal entries equal to λ_1.

This takes care of the generalized eigenspace $W_1 = E_\infty(\lambda_1)$. Perform the same process for each generalized eigenspace W_i, obtaining a basis B_i with $[T_i]_{B_i}$ in Jordan Canonical Form, for each i. Then, if $B = B_1 \cup \ldots \cup B_m$, we have that $A = [T]_B$ is a matrix in Jordan Canonical Form.   □

(Observe in the proof that it was crucial that we started with the *longest* chain when working with each eigenvalue.)

Again, we proved a more precise result than we have stated. Let us state that more precise result now.

Theorem 2.5.6. *Let V be a vector space and let $T : V \longrightarrow V$ be a linear transformation with characteristic polynomial $c_T(x) = (x - \lambda_1)^{k_1}(x - \lambda_2)^{k_2} \cdots (x - \lambda_m)^{k_m}$. and minimum polynomial $m_T(x) = (x - \lambda_1)^{j_1}(x - \lambda_2)^{j_2} \cdots (x - \lambda_m)^{j_m}$. Then V has a basis B in which $J = [T]_B$ is a matrix in Jordan Canonical Form*

$$J = \begin{bmatrix} J_1 & & & & & \\ & J_2 & & & 0 & \\ & & J_3 & & & \\ & & & \ddots & & \\ & 0 & & & J_{p-1} & \\ & & & & & J_p \end{bmatrix},$$

where $J_1, \ldots, J_p$ are all Jordan blocks.

Furthermore, for each i:

(1) the sum of the sizes of the Jordan blocks with diagonal entries λ_i is equal to $k_i = $ alg-mult(λ_i);

(2) the largest Jordan block with diagonal entries λ_i is j_i-by-j_i where $j_i = $ max-ind(λ_i); and

(3) the number of Jordan blocks with diagonal entries λ_i is equal to geom-mult(λ_i).

Proof. Parts (1) and (2) are clear from the proof of Theorem 2.5.5.

As for part (3), we note that the vectors at the bottom of each of the chains for the eigenvalue λ_i form a basis for the eigenspace $E_1(\lambda_i)$. But the number of these vectors is the number of blocks, and the dimension of $E_1(\lambda_i)$ is the geometric multiplicity of λ_i.   □

Remark 2.5.7. Once we have that $\mathcal{T}$ has a Jordan Canonical Form, it is natural to ask whether its Jordan Canonical Form is unique. There is one clear indeterminacy, as there is no a priori order to the blocks, so they can be reordered at will. This turns out to be the only indeterminacy. We will defer the formal statement of this to Theorem 3.2.9, when we will be in a position to easily prove it. (Although we have the tools to prove it now, doing so would involve us in some repetition, so it is better to wait.)   ◇

Theorem 2.5.6 leads us to the following definition.

Definition 2.5.8. Let V be a vector space and let $\mathcal{T} : V \longrightarrow V$ be a linear transformation. A basis $\mathcal{B}$ of V is a *Jordan basis* of V with respect to $\mathcal{T}$ if $J = [\mathcal{T}]_{\mathcal{B}}$ is a matrix in Jordan Canonical Form.   ◇

Remark 2.5.9. As opposed to Remark 2.5.7, which states that the Jordan Canonical Form of $\mathcal{T}$ is essentially unique, there is never any uniqueness result for a Jordan basis $\mathcal{B}$ of V with respect to $\mathcal{T}$. There are always infinitely many such bases $\mathcal{B}$. (Again, it is convenient to delay illustrating this phenomenon.)   ◇

We conclude this section by restating Theorem 2.5.6 in terms of matrices.

Theorem 2.5.10. *Let A be an n-by-n matrix with characteristic polynomial $c_T(x) = (x - \lambda_1)^{k_1}(x - \lambda_2)^{k_2} \cdots (x - \lambda_m)^{k_m}$ and minimum polynomial $m_T(x) = (x - \lambda_1)^{j_1}(x - \lambda_2)^{j_2} \cdots (x - \lambda_m)^{j_m}$. Then A is similar to a matrix J in Jordan Canonical Form,*

$$
J = \begin{bmatrix} J_1 & & & & & \\ & J_2 & & 0 & & \\ & & J_3 & & & \\ & & & \ddots & & \\ & 0 & & & J_{p-1} & \\ & & & & & J_p \end{bmatrix},
$$

where $J_1, \ldots, J_p$ are all Jordan blocks.

Furthermore, for each i:

(1) the sum of the sizes of the Jordan blocks with diagonal entries λ_i is equal to $k_i = \text{alg-mult}(\lambda_i)$;

(2) the largest Jordan block with diagonal entries λ_i is j_i-by-j_i where $j_i = \text{max-ind}(\lambda_i)$; and

(3) the number of Jordan blocks with diagonal entries λ_i is equal to $\text{geom-mult}(\lambda_i)$.

Proof. Immediate from Theorem 2.5.6.   □

2.6   EXERCISES

In of the following exercises, you are given some of the following information about a matrix A: Its characteristic polynomial $c_A(x)$, its minimum polynomial $m_A(x)$, its eigenvalues, and the algebraic and geometric multiplicities and maximum index of each eigenvalue. In each case, find all possible Jordan Canonical Forms J of A consistent with the given information, and for each such, find the rest of the information.

1. $c_A(x) = (x - 2)(x - 3)(x - 5)(x - 7)$.

2. A is 4-by-4, $m_A(x) = x - 1$.

3. $c_A(x) = (x - 2)^3(x - 3)^2$.

4. $c_A(x) = (x - 7)^5$.

5. $c_A(x) = (x - 3)^4(x - 5)^4$, $m_A(x) = (x - 3)^2(x - 5)^2$.

6. $c_A(x) = x(x - 4)^7$, $m_A(x) = x(x - 4)^3$.

7. $c_A(x) = x^2(x - 2)^4(x - 6)^2$, max-ind$(0) = 1$, max-ind$(2) = 2$, max-ind$(6) = 2$.

8. $c_A(x) = (x - 3)^3(x - 4)^3$, max-ind$(3) = 2$, max-ind$(4) = 1$.

9. $c_A(x) = (x - 3)^4(x - 8)^2$, geom-mult$(3) = 2$, geom-mult$(8) = 1$.

10. $c_A(x) = (x - 7)^5$, geom-mult$(7) = 3$.

11. A is 6-by-6, $m_A(x) = (x - 1)^4(x - 4)$.

12. A is 6-by-6, $m_A(x) = (x - 2)^2(x - 3)$.

13. alg-mult$(1) = 5$, geom-mult$(1) = 3$.

14. alg-mult$(2) = 6$, geom-mult$(2) = 3$.

15. alg-mult$(3) = 5$, max-ind$(3) = 2$.

16. alg-mult$(4) = 6$, max-ind$(4) = 3$.

17. alg-mult$(8) = 6$, geom-mult$(8) = 3$, max-ind$(8) = 4$.

18. alg-mult$(9) = 6$, geom-mult$(4) = 4$, max-ind$(4) = 2$.

19. (a) Suppose A is an n-by-n matrix with $n \leq 3$. Show that A is determined up to similarity by its characteristic polynomial and its minimum polynomial.

   (b) Show by example that this is not true if $n = 4$.

20. (a) Suppose A is an n-by-n matrix with $n \leq 3$. Show that A is determined up to similarity by its characteristic polynomial and the geometric multiplicities of each of its eigenvalues.

   (b) Show by example that this is not true if $n = 4$.

21. (a) Suppose A is an n-by-n matrix with $n \leq 6$. Show that A is determined up to similarity by its characteristic polynomial, its minimum polynomial, and the geometric multiplicities of each of its eigenvalues.

   (b) Show by example that this is not true if $n = 7$.

22. For $n \geq 2$ and $b \neq 0$, let $M = M_n(a, b)$ be the n-by-n matrix all of whose diagonal entries are equal to a and all of whose off-diagonal entries are equal to b.

(a) Show that $a - b$ is an eigenvalue of M of geometric multiplicity $n - 1$.

(b) Show that $a + (n - 1)b$ is an eigenvalue of M of geometric multiplicity 1.

(c) Use the results of parts (a) and (b) to find the characteristic polynomial $c_M(x)$ and the minimum polynomial $m_M(x)$.

(d) Find the Jordan Canonical Form J of M and an invertible matrix P with $M = PJP^{-1}$.

CHAPTER 3

An Algorithm for Jordan Canonical Form and Jordan Basis

Our objective in this chapter is to develop an algorithm for finding the *JCF* of a linear transformation, assuming that one can find its eigenvalues, i.e., that one can factor its characteristic (or minimum) polynomial. We will further develop the algorithm to find a Jordan basis as well. We begin this development by defining the *eigenstructure picture* (or *ESP*) of a linear transformation, which represents its *JCF* in an way that is easy to visualize. Then we develop the algorithm per se, and illustrate it with many examples.

3.1 THE *ESP* OF A LINEAR TRANSFORMATION

In this section, we present the eigenstructure picture, or *ESP*, of a linear transformation $\mathcal{T} : V \longrightarrow V$.

Part of our discussion will be general, but we shall also proceed partly by example, as to proceed in complete generality would lead to a blizzard of subscripts, which would only confuse the issue. But our examples should be general enough to make the situation clear.

Here is the basic element of an *ESP*:

Construction 3.1.1. Let $\mathcal{T} : V \longrightarrow V$ be a linear transformation and let λ be an eigenvalue of $\mathcal{T}$. Suppose that $\mathcal{T}$ has a chain of generalized eigenvectors (see Definition 2.5.1) $\{v_1, \ldots, v_k\}$ of length k associated to the eigenvector λ. In this situation, we have the following "labelled picture" in Figure 3.1.

By way of explanation, each integer on the left denote the index of the relevant generalized eigenvector. Each node represents a generalized eigenvector of that index, so in this picture, index is synonymous with height (with the lowest level being at height 1). The solid links between the nodes indicate that these generalized eigenvectors are all part of a chain. The complex number at the bottom is the associated eigenvalue. Finally, each node is labelled with the corresponding generalized eigenvector.

In this situation, we also have the "unlabelled picture" (or simply "picture"), which is obtained from the labelled picture simply by deleting all the labels. ◇

We will often think of Construction 3.1.1 the other way around. That is, we will think of beginning with the picture and obtaining the labelled picture from it by adding labels to all the nodes.

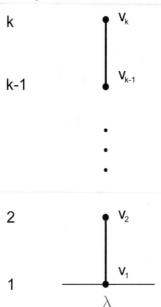

Figure 3.1.

Construction 3.1.2. Let $T : V \longrightarrow V$ be a linear transformation. The *eigenstructure picture*, or *ESP*, of T is the picture obtained by writing down all the pictures of its various chains next to each other. (We only write the indices once, to avoid duplication.)

The *labelled eigenstructure picture*, or *ℓESP*, of T is the picture obtained by writing down all the labelled pictures of its various chains next to each other (again, with the indices only written once, to avoid duplication).

We call a node in a column associated to an eigenvalue λ a λ-node.   ◇

Example 3.1.3. (1) Let $T : V \longrightarrow V$ be a linear transformation with *ESP* the picture in Figure 3.1. Then the *JCF* J of T is a single k-by-k Jordan block,

$$
J = \begin{bmatrix}
\lambda & 1 \\
& \lambda & 1 \\
& & \lambda & 1 \\
& & & \ddots & \ddots \\
& & & & \lambda & 1 \\
& & & & & \lambda
\end{bmatrix}.
$$

Furthermore, if $\mathcal{T}$ has ℓESP the labelled picture in Figure 3.1, then $J = [\mathcal{T}]_\mathcal{B}$, i.e., $\mathcal{B}$ is a Jordan basis of V with respect to $\mathcal{T}$, where $\mathcal{B}$ is the basis $\mathcal{B} = \{v_1, \ldots, v_k\}$ of V.

(2) Let $\mathcal{T} : V \longrightarrow V$ be a linear transformation with ℓESP

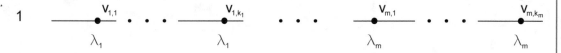

Figure 3.2.

In this case, every generalized eigenvector of $\mathcal{T}$ is an eigenvector of $\mathcal{T}$, so $\mathcal{T}$ is diagonalizable and its *JCF* is the diagonal matrix

$$
J = \begin{bmatrix}
\lambda_1 & & & & & \\
& \ddots & & & & \\
& & \lambda_1 & & & \\
& & & \ddots & & \\
& & & & \lambda_m & \\
& & & & & \ddots & \\
& & & & & & \lambda_m
\end{bmatrix}.
$$

Furthermore, if $\mathcal{T}$ has ℓESP the labelled picture in Figure 3.2, then $J = [\mathcal{T}]_\mathcal{B}$, i.e., $\mathcal{B}$ is a Jordan basis of V with respect to $\mathcal{T}$, where $\mathcal{B}$ is the basis $\mathcal{B} = \{v_{1,1}, \ldots, v_{1,k_1}, \ldots, v_{m,1}, \ldots, v_{m,k_m}\}$ of V.

(3) Let $\mathcal{T} : V \longrightarrow V$ be a linear transformation with ℓESP (Figure 3.3). Then $\mathcal{T}$ has Jordan Canonical Form

$$
J = \begin{bmatrix}
6 & 1 & & & & & & & & \\
& 6 & 1 & & & & & & & \\
& & 6 & 1 & & & & & & \\
& & & 6 & & & & & & \\
& & & & 6 & 1 & & & & \\
& & & & & 6 & & & & \\
& & & & & & 7 & 1 & & \\
& & & & & & & 7 & & \\
& & & & & & & & 7 & \\
& & & & & & & & & 7
\end{bmatrix}.
$$

Note that J has a 4-by-4 block and a 2-by-2 block for the eigenvalue 6, and a 2-by-2 block and two 1-by-1 blocks for the eigenvalue 7.

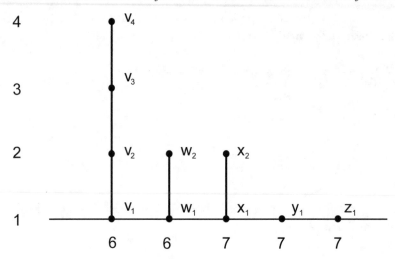

Figure 3.3.

Also, $\mathcal{B} = \{v_1, v_2, v_3, v_4, w_1, w_2, x_1, x_2, y_1, z_1\}$ is a Jordan basis of V with respect to $\mathcal{T}$.

(4) Let $\mathcal{T} : V \longrightarrow V$ be a linear transformation with ℓESP

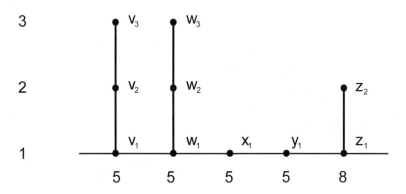

Figure 3.4.

Then $\mathcal{T}$ has Jordan Canonical Form

$$J = \begin{bmatrix} 5 & 1 & & & & & & & & \\ & 5 & 1 & & & & & & & \\ & & 5 & & & & & & & \\ & & & 5 & 1 & & & & & \\ & & & & 5 & 1 & & & & \\ & & & & & 5 & & & & \\ & & & & & & 5 & & & \\ & & & & & & & 5 & & \\ & & & & & & & & 8 & 1 \\ & & & & & & & & & 8 \end{bmatrix}.$$

Note that J has two 3-by-3 blocks and two 1-by-1 blocks for the eigenvalue 5, and a 2-by-2 block for the eigenvalue 8.

Also, $\mathcal{B} = \{v_1, v_2, v_3, w_1, w_2, w_3, x_1, y_1, z_1, z_2\}$ is a Jordan basis of V with respect to $\mathcal{T}$.

(5) Let $\mathcal{T} : V \longrightarrow V$ be a linear transformation with ℓESP

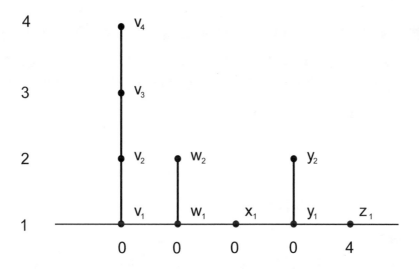

Figure 3.5.

Then $\mathcal{T}$ has Jordan Canonical Form

$$J = \begin{bmatrix} 0 & 1 & & & & & & & & \\ & 0 & 1 & & & & & & & \\ & & 0 & 1 & & & & & & \\ & & & 0 & & & & & & \\ & & & & 0 & 1 & & & & \\ & & & & & 0 & & & & \\ & & & & & & 0 & & & \\ & & & & & & & 1 & 1 & \\ & & & & & & & & 1 & \\ & & & & & & & & & 4 \end{bmatrix}.$$

Note that J has a 4-by-4 block, a 2-by-2 block, and a 1-by-1 block for the eigenvalue 0, a 2-by-2 block for the eigenvalue 1, and a 1-by-1 block for the eigenvalue 4.

Also, $\mathcal{B} = \{v_1, v_2, v_3, v_4, w_1, w_2, x_1, y_1, y_2, z_1\}$ is a Jordan basis of V with respect to $\mathcal{T}$.

$\diamond$

From Example 3.1.3, it should be clear how to translate between the *ESP* of a linear transformation and its *JCF*, and that each of these determines the other. It should also be clear how to translate between an ℓ*ESP* of a linear transformation and its *JCF* together with a Jordan basis, and that each of these determines the other.

Note that we may reorder the columns in the *ESP* of a linear transformation, and that corresponds to reordering the blocks in its *JCF*.

3.2   THE ALGORITHM FOR JORDAN CANONICAL FORM

In this section, we develop our algorithm. Actually, what we will develop is an algorithm for determining the *ESP* of a linear transformation. In light of the previous section, that determines its *JCF*.

We fix a linear transformation $\mathcal{T} : V \longrightarrow V$.

Let us reestablish some notation. Recall Definition 2.1.14 and Definition 2.1.5 (both restated in terms of a linear transformation): For an eigenvalue λ of $\mathcal{T}$, and a fixed integer j, we let $E_j(\lambda)$ be the set of vectors

$$E_j(\lambda) = \{v \mid (\mathcal{T} - \lambda\mathcal{I})^j v = 0\} = \mathrm{Ker}((\mathcal{T} - \lambda\mathcal{I})^j).$$

Also, we let $E_\infty(\lambda)$ be the set of vectors

$$E_\infty(\lambda) = \{v \mid (\mathcal{T} - \lambda\mathcal{I})^j v = 0 \text{ for some } j\} = \{v \mid v \text{ is in } E_j(\lambda) \text{ for some } j\}.$$

($E_\infty(\lambda)$ is the generalized eigenspace of λ.)

We have

$$E_1(\lambda) \subseteq E_2(\lambda) \subseteq \ldots \subseteq E_\infty(\lambda),$$

with each of these being a $\mathcal{T}$-invariant subspace of V.

We then set

$$d_j(\lambda) = \dim E_j(\lambda) \text{ for } j = 1, 2, \ldots, \infty.$$

Observation 3.2.1. We note that the labels on the nodes give linearly independent vectors, and that the λ-nodes at height j or below form a basis for $E_j(\lambda)$, i.e.,

$$d_j(\lambda) = \text{ the number of } \lambda\text{-nodes at height } j \text{ or below in the } ESP \text{ of } \mathcal{T}.$$

$\diamond$

Remark 3.2.2. Observation 3.2.1 illustrates the conclusion of Proposition 2.1.19: If $j_{\max} = \text{max-ind}(\lambda)$ then $d_\infty(\lambda) = d_{j_{\max}}(\lambda)$ (as all the nodes in the ESP of $\mathcal{T}$ are at height $j_{\max}$ or below) and furthermore that $j = j_{\max}$ is the smallest value of j such that $d_{j+1}(\lambda) = d_j(\lambda)$ (as for any smaller value of j there are nodes above height j).

We also recall one of the conclusions of Theorem 2.5.6: $d_\infty(\lambda) = \text{alg-mult}(\lambda)$, from which we see that $j = j_{\max}$ is the smallest value of j such that $d_j(\lambda) = \text{alg-mult}(\lambda)$. $\diamond$

Observation 3.2.3. Let us set $d_j^{\text{ex}}(\lambda)$ equal to the number of nodes that are at height (exactly) j. Then we see that

$$d_j^{\text{ex}}(\lambda) = d_j(\lambda) - d_{j-1}(\lambda) \text{ for } j > 1, \text{ and } d_j^{\text{ex}}(\lambda) = d_j(\lambda) \text{ for } j = 1.$$

$\diamond$

Remark 3.2.4. We are trying to find the ESP of $\mathcal{T}$, i.e., to find the positions of all the nodes in its ESP. But in order to do that, it certainly suffices to find the positions of the nodes that are at the top of the chains (since, once we have those, we may simply fill in the rest of the chains). Let us think of those nodes as "new" nodes and the other nodes as "old" nodes. We use this language as we think of working from the top of the ESP down, so the "new" nodes at height j are ones that first show up at height j, while the "old" nodes are parts of chains that are continuing down from height $j + 1$ or above. $\diamond$

Observation 3.2.5. Let us set $d_j^{\text{new}}(\lambda)$ equal to the number of new nodes that are at height j. Then we see that

$$d_j^{\text{new}}(\lambda) = d_j^{\text{ex}}(\lambda) \text{ for } j = j_{\max}, \text{ and } d_j^{\text{new}}(\lambda) = d_j^{\text{ex}}(\lambda) - d_{j+1}^{\text{ex}}(\lambda) \text{ for } j < j_{\max}.$$

$\diamond$

Assembling these observations and remarks gives us our algorithm for determining the ESP, or equivalently the JCF, of $\mathcal{T}$.

Algorithm 3.2.6. *(Algorithm for ESP/JCF) For each eigenvalue λ of $\mathcal{T}$:*

(1) For $j = 1, 2, \ldots$ find $E_j(\lambda) = \mathrm{Ker}(\mathcal{T} - \lambda \mathcal{I})$, and let $d_j(\lambda) = \dim E_j(\lambda)$ be its dimension. Stop when $j = j_{\max} = \mathrm{max\text{-}ind}(\lambda)$, a value that is attained when $d_j(\lambda) = \mathrm{alg\text{-}mult}(\lambda)$, or equivalently when $d_{j+1}(\lambda) = d_j(\lambda)$.

(2) For $j = j_{\max}$ down to 1 compute $d_j^{\mathrm{ex}}(\lambda)$ by:

$$d_j^{\mathrm{ex}}(\lambda) = d_j(\lambda) - d_{j-1}(\lambda) \text{ for } j > 1,$$
$$d_j^{\mathrm{ex}}(\lambda) = d_j(\lambda) \text{ for } j = 1.$$

(3) For $j = j_{\max}$ down to 1 compute $d_j^{\mathrm{new}}(\lambda)$ by:

$$d_j^{\mathrm{new}}(\lambda) = d_j^{\mathrm{ex}}(\lambda) \text{ for } j = j_{\max},$$
$$d_j^{\mathrm{new}}(\lambda) = d_j^{\mathrm{ex}}(\lambda) - d_{j+1}^{\mathrm{ex}}(\lambda) \text{ for } j < j_{\max}.$$

Then the columns of the ESP of $\mathcal{T}$ associated to the eigenvalue λ consists of $d_j^{\mathrm{new}}(\lambda)$ chains of length j, for each j, or equivalently the portion of the JCF of $\mathcal{T}$ associated to the eigenvalue λ consists of $d_j^{\mathrm{new}}(\lambda)$ j-by-j blocks, for each j.

Remark 3.2.7. Simply by counting nodes, we see that, for each eigenvalue λ of $\mathcal{T}$,

$$\sum_{j=1}^{j_{\max}} j d_j^{\mathrm{new}}(\lambda) = \mathrm{alg\text{-}mult}(\lambda).$$

$\diamond$

With this algorithm in hand, we can easily obtain a uniqueness result for Jordan Canonical Form.

Theorem 3.2.8. *Let $\mathcal{T} : V \longrightarrow V$ be a linear transformation. Then the Jordan Canonical Form of $\mathcal{T}$ is unique up to the order of the blocks.*

Proof. Given $\mathcal{T}$, its eigenvalues are uniquely determined; for any eigenvalue λ of $\mathcal{T}$, the spaces $E_j(\lambda)$ are uniquely determined, and so the integers $d_j(\lambda)$ are uniquely determined; then the integers $d_j^{\mathrm{ex}}(\lambda)$ are uniquely determined; and then the integers $d_j^{\mathrm{new}}(\lambda)$ are uniquely determined. Thus the ESP of $\mathcal{T}$ is uniquely determined up to the order of the columns; correspondingly, the JCF of $\mathcal{T}$ is uniquely determined up to the order of the blocks. $\square$

To give the reader a feel for how Algorithm 3.2.6 works, we shall first see what it gives in the cases of Example 3.1.3, where we already know the *ESP* (or *JCF*). Later in this chapter, we will apply it to determine *ESP* (or *JCF*) in cases where we do not know it already.

To simplify notation in applying Algorithm 3.2.6, once we have fixed an eigenvalue λ, we shall denote $d_j(\lambda)$ simply by d_j, etc.

Example 3.2.9. (1) Let $\mathcal{T}$ be as in Example 3.1.3 (1). Then

$$d_j(\lambda) = j \text{ for } j = 1, \ldots, k.$$

Since $d_k = k = \text{alg-mult}(\lambda)$ but $d_j < k = \text{alg-mult}(\lambda)$ for $j < k$, we have that $j_{\max} = k$. Then

$$d_j^{\text{ex}}(\lambda) = j - (j - 1) = 1 \text{ for } j = k, k - 1, \ldots, 1,$$

and then

$$d_k^{\text{new}}(\lambda) = 1 - 0 = 1 \text{ and } d_j^{\text{new}}(\lambda) = 1 - 1 = 0 \text{ for } j = k - 1, \ldots, 1.$$

Thus the *ESP* of $\mathcal{T}$ has 1 chain of length k associated to the eigenvalue λ.

(2) Let $\mathcal{T}$ be as in Example 3.1.3 (2). Then for each eigenvalue λ_i of $\mathcal{T}$,

$$d_1(\lambda_i) = k_i.$$

Since $k_i = \text{alg-mult}(\lambda_i)$, we have that $j_{\max}(\lambda_i) = 1$. Then

$$d_1^{\text{ex}}(\lambda_i) = k_i,$$

and then

$$d_1^{\text{new}}(\lambda_i) = k_i.$$

Thus the *ESP* of $\mathcal{T}$ has k_i chains of length 1 associated to the eigenvalue λ_i.

(3) Let $\mathcal{T}$ be as in Example 3.1.3 (3).
For the eigenvalue 6 of $\mathcal{T}$,

$$d_1 = 2, \ d_2 = 4, \ d_3 = 5, \ d_4 = 6,$$

and $j_{\max} = 4$. Then

$$d_4^{\text{ex}} = 6 - 5 = 1, \ d_3^{\text{ex}} = 5 - 4 = 1, \ d_2^{\text{ex}} = 4 - 2 = 2, \ d_1^{\text{ex}} = 2,$$

and then

$$d_4^{\text{new}} = 1, \ d_3^{\text{new}} = 1 - 1 = 0, \ d_2^{\text{new}} = 2 - 1 = 1, \ d_1^{\text{new}} = 2 - 2 = 0.$$

Thus the *ESP* of $\mathcal{T}$ has 1 chain of length 4 and 1 chain of length 2 associated to the eigenvalue 6.
For the eigenvalue 7 of $\mathcal{T}$,

$$d_1 = 3, \ d_2 = 4,$$

and $j_{\max} = 2$. Then

$$d_2^{\text{ex}} = 4 - 3 = 1, \ d_1^{\text{ex}} = 3,$$

and then

$$d_2^{\text{new}} = 1, \ d_1^{\text{new}} = 3 - 1 = 2.$$

Thus the *ESP* of $\mathcal{T}$ has 1 chain of length 2 and 2 chains of length 1 associated to the eigenvalue 7.

(3) Let $\mathcal{T}$ be as in Example 3.1.3 (4).
For the eigenvalue 5 of $\mathcal{T}$,

$$d_1 = 4, \ d_2 = 6, \ d_3 = 8,$$

and $j_{\max} = 3$. Then

$$d_3^{\text{ex}} = 8 - 6 = 2, \ d_2^{\text{ex}} = 6 - 4 = 2, \ d_1^{\text{ex}} = 4,$$

and then

$$d_3^{\text{new}} = 2, \ d_2^{\text{new}} = 2 - 2 = 0, \ d_1^{\text{new}} = 4 - 2 = 2.$$

Thus the *ESP* of $\mathcal{T}$ has 2 chains of length 3 and 2 chains of length 1 associated to the eigenvalue 5.
For the eigenvalue 8 of $\mathcal{T}$,

$$d_1 = 1, \ d_2 = 2,$$

and $j_{\max} = 2$. Then

$$d_2^{\text{ex}} = 2 - 1 = 1, \ d_1^{\text{ex}} = 1,$$

and then

$$d_2^{\text{new}} = 1, \ d_1^{\text{new}} = 1 - 1 = 0.$$

Thus the *ESP* of $\mathcal{T}$ has 1 chain of length 2 associated to the eigenvalue 8.

(3) Let $\mathcal{T}$ be as in Example 3.1.3 (5).
For the eigenvalue 0 of $\mathcal{T}$,

$$d_1 = 3, \ d_2 = 5, \ d_3 = 6, \ d_4 = 7,$$

and $j_{\max} = 4$. Then

$$d_4^{\text{ex}} = 7 - 6 = 1, \ d_3^{\text{ex}} = 6 - 5 = 1, \ d_2^{\text{ex}} = 5 - 3 = 2, \ d_1^{\text{ex}} = 3,$$

and then

$$d_4^{\text{new}} = 1, \ d_3^{\text{new}} = 1 - 1 = 0, \ d_2^{\text{new}} = 2 - 1 = 1, \ d_1^{\text{new}} = 3 - 2 = 1.$$

Thus the *ESP* of $\mathcal{T}$ has 1 chain of length 4, 1 chain of length 2, and 1 chain of length 1 associated to the eigenvalue 0.
For the eigenvalue 1 of $\mathcal{T}$,

$$d_1 = 1, \ d_2 = 2,$$

and $j_{\max} = 2$. Then

$$d_2^{\text{ex}} = 2 - 1 = 1, \ d_1^{\text{ex}} = 1,$$

and then

$$d_2^{\text{new}} = 1, \ d_1^{\text{new}} = 1 - 1 = 0.$$

Thus the *ESP* of $\mathcal{T}$ has 1 chain of length 2 associated to the eigenvalue 1.

For the eigenvalue 4 of $\mathcal{T}$,

$$d_1 = 1,$$

and $j_{\max} = 1$. Then

$$d_1^{\mathrm{ex}} = 1,$$

and then

$$d_1^{\mathrm{new}} = 1.$$

Thus the *ESP* of $\mathcal{T}$ has 1 chain of length 1 associated to the eigenvalue 4.   ◇

3.3   THE ALGORITHM FOR A JORDAN BASIS

We continue to fix a linear transformation $\mathcal{T} : V \longrightarrow V$.

In this section, we further develop our algorithm in order to obtain a Jordan basis of V with respect to $\mathcal{T}$. We have already developed an algorithm for determining the *ESP* of a linear transformation. Now we will extend our algorithm in order to determine labels on its nodes, and hence a *ℓESP* of $\mathcal{T}$. We can then directly read off a Jordan basis from the labels on the nodes.

We have been careful in our language. As we have already remarked, and as we shall see concretely below, there is no uniqueness whatsoever for a Jordan basis, so we will be finding *a*, not *the*, Jordan basis of V with respect to $\mathcal{T}$.

We suppose that we have already found the *ESP* of $\mathcal{T}$.

We will be working one eigenvalue at a time. For each eigenvalue λ, we will be working from the top down from the top down, that is, starting with $j_{\max} = \text{max-ind}(\lambda)$ and ending with $j = 1$. Recall Remark 3.2.4: At any stage, that is, for any given value of j, we think of the nodes that are at the top of chains of height j as "new" nodes, since we see them for the first time at this stage. The other nodes at height j are parts of chains we have already seen, and so we think of them as "old" nodes. Note that when we first start, with $j = j_{\max}$, every node at this height is a new node. Our task is then to find the labels on the new nodes at each stage.

To clarify this, let us look back at our examples.

First, consider Example 3.1.3 (3). For the eigenvalue 6, we start with $j = 4$ and work our way down. For $j = 4$, we have one new node, v_4. For $j = 3$, we have one old node, v_3, and no new nodes. For $j = 2$, we have one old node, v_2, and one new node, w_2. For $j = 1$, we have two old nodes, v_1 and w_1, and no new nodes. For the eigenvalue 7, we start with $j = 2$ and work our way down. For $j = 2$, we have one new node, x_2. For $j = 1$, we have one old node, x_1, and two new nodes, y_1 and z_1.

Next, consider Example 3.1.3 (4). For the eigenvalue 5, we start with $j = 3$ and work our way down. For $j = 3$, we have two new nodes, v_3 and w_3. For $j = 2$, we have two old nodes, v_2 and w_2, and no new nodes. For $j = 1$, we have two old nodes, v_1 and w_1, and two new nodes, x_1 and y_1. For the eigenvalue 8, we start with $j = 2$ and work our way down. For $j = 2$, we have one new node, z_2. For $j = 1$, we have one old node, z_1, and no new nodes.

Last, consider Example 3.1.3 (5). For the eigenvalue 0, we start with $j = 4$ and work our way down. For $j = 4$, we have one new node, v_4. For $j = 3$, we have one old node, v_3, and no new nodes. For $j = 2$, we have one old node, v_2, and one new node, w_2. For $j = 1$, we have two old nodes, v_1 and w_1, and one new node x_1. For the eigenvalue 1, we start with $j = 2$ and work our way down. For $j = 2$, we have one new node, y_2. For $j = 1$, we have one old node, y_1, and no new nodes. For the eigenvalue 4, we start and end with $j = 1$, and for $j = 1$ we have one new node z_1.

Now we get to work.

We begin by observing that we need only determine the labels on the new nodes, i.e., the labels on the nodes that are at the top of chains. For, once we have determined these, the remaining labels, i.e., the labels on the old nodes, are then determined, as we see from Remark 2.5.2.

We proceed one eigenvalue at a time, so we fix an eigenvalue λ of $\mathcal{T}$. As we did in the case of our algorithm for finding the *ESP* of $\mathcal{T}$, we shall once again proceed from the top down.

We again set $\mathcal{S} = \mathcal{T} - \lambda\mathcal{I}$, for simplicity.

Let us recall our notation. For a fixed integer j, $E_j(\lambda)$ is the set of vectors

$$E_j(\lambda) = \{v \mid (\mathcal{T} - \lambda\mathcal{I})^j v = 0\} = \mathrm{Ker}((\mathcal{T} - \lambda\mathcal{I})^j) = \mathrm{Ker}(\mathcal{S}^j).$$

Since we are fixing λ, we shall abbreviate $E_j(\lambda)$ by E_j, etc.

Now suppose we are at height j. The E_j has a basis consisting of the labels on all the nodes at height at most j. These fall into three classes:

(1): The labels on all the nodes at height at most $j - 1$. These form a basis for E_{j-1}.

(2): The labels on the old nodes at height j. These form a basis for a subspace of E_j which we will denote by A_j. (Here "A" stands for alt, the German word for old. We cannot use "O" as that would be too confusing typographically.)

(3): The labels on the new nodes at height j. These form a basis for a subspace of E_j which we will denote by N_j. (Here "N" stands for neu or new, take your pick.)

Thus we see that $E_j = E_{j-1} \oplus A_j \oplus N_j$. Let us also set $F_j = A_j \oplus N_j$, so that F_j is the subspace of E_j with basis the nodes at height exactly j.

What we are trying to find, of course, are the labels in (3). What do we know at this stage? We can certainly find the subspaces E_j and E_{j-1} without any trouble, and indeed bases for these subspaces. (The trouble is in finding the right kind of bases for these subspaces.) Also, we know A_j, as $A_j = \mathcal{S}(F_{j+1})$, and we found F_{j+1} at the previous stage. (If we are at the top, there is no F_{j+1}.) But not only do we know A_j, we know a basis for A_j, since we will have found a basis for F_{j+1} at an earlier stage, and we can obtain a basis for A_j by applying $\mathcal{S}$ to each element of a basis for F_{j+1}.

But now we know how to find a basis for N_j: Find a basis $\mathcal{B}_1$ for E_{j-1} (any basis). Let $\mathcal{B}_2$ be the basis of A_j obtained by applying $\mathcal{S}$ to the basis of F_{j+1} we obtained at the previous stage. Extend $\mathcal{B}_1 \cup \mathcal{B}_2$ to a basis $\mathcal{B} = \mathcal{B}_1 \cup \mathcal{B}_2 \cup \mathcal{B}_3$ of E_j. Then $\mathcal{B}_3$ is a basis of N_j, and we label the new nodes at height j with the vectors in $\mathcal{B}_3$.

This finishes us at height j, except that to get ready for height $j - 1$ we observe that $\mathcal{B}_2 \cup \mathcal{B}_3$ is a basis for $A_j \oplus N_j = F_j$. (If $j = 1$, we don't have to bother with this.)

We now summarize the algorithm we have just developed to find a ℓESP of $\mathcal{T}$, or, equivalently, a Jordan basis for V with respect to $\mathcal{T}$. It is easiest to state this algorithm if we adopt the convention that if $j = j_{\max}$, then $F_{j+1} = \{0\}$ with empty basis, and that if $j = 1$, then $E_{j-1} = \{0\}$ with empty basis.

Algorithm 3.3.1. *(Algorithm for ℓESP/Jordan basis) For each eigenvalue λ of $\mathcal{T}$:*

(1) For $j = 1, 2, \ldots, j_{\max} = \text{max-ind}(\lambda)$, find $E_j(\lambda) = \text{Ker}(\mathcal{T} - \lambda \mathcal{I})$.

(2) For $j = j_{\max}$ down to 1:

(a) Find a basis $\mathcal{B}_1$ of E_{j-1}.

(b) Apply $\mathcal{S} = \mathcal{T} - \lambda \mathcal{I}$ to the basis of F_{j+1} obtained in the previous stage to obtain a basis $\mathcal{B}_2$ of A_j. Label the old nodes at height j with this basis, in the same order as the basis of F_{j+1}.

(c) Extend $\mathcal{B}_1 \cup \mathcal{B}_2$ to a basis $\mathcal{B} = \mathcal{B}_1 \cup \mathcal{B}_2 \cup \mathcal{B}_3$ of E_j. Then $\mathcal{B}_3$ is a basis of N_j. Label the new nodes at height j with the vectors in $\mathcal{B}_3$.

(d) Let $F_j = A_j \oplus N_j$. (Note we have just given F_j a basis, namely the labels in (b) and (c).)

Remark 3.3.2. Note that the bases $\mathcal{B}_1$ which we arbitrarily chose at any stage were just used to help us find the bases $\mathcal{B}_3$, the ones we are really interested in. These bases $\mathcal{B}_1$ will not, in general, appear in the ℓESP. In step 2 (c), we have to extend $\mathcal{B}_2 \cup \mathcal{B}_3$ to a basis $\mathcal{B}$ of E_j. How do we go about doing that? In practice, we proceed as follows: First, find a basis $\mathcal{C}_1$ of E_j. Then there is a subset $\mathcal{C}_2$ of $\mathcal{C}_1$ such that $\mathcal{B}_1 \cup \mathcal{C}_2$ is a basis of E_j. Then our desired basis $\mathcal{B}_3 = \mathcal{C}_3$ is a further subset of $\mathcal{C}_2$ such that $\mathcal{B}_1 \cup \mathcal{B}_2 \cup \mathcal{C}_3$ is a basis of E_j. But again neither $\mathcal{C}_1$ nor $\mathcal{C}_2$ appears in the ℓESP.   ◇

Remark 3.3.3. F_j is a complement of E_{j-1} in E_j. Since E_{j-1} consists of all generalized eigenvectors of index at most $j - 1$, and the 0 vector, that implies that every nonzero vector in F_j is a generalized eigenvector of index j. But that is *not* saying that F_j contains every generalized eigenvector of index j, and that is in general false. Indeed, for $j > 1$, $S = \{$generalized eigenvectors of index $j\} \cup \{0\}$ is *not* a subspace of V, as we can see from the following argument: Let v be a generalized eigenvector of index j, so v is in S. Let v' be a generalized eigenvector of index $j - 1$. Then $w = v + v'$ is also a generalized eigenvector of index j, so w is also in S. But $v' = w - v$ is not in S, so S is not a subspace. This fact accounts for part of the care that we had to take in finding N_j.   ◇

Remark 3.3.4. There is one more point we have to deal with to ensure that our algorithm works. The labels on the nodes in F_j are independent. For $j > 1$, we need to know that the labels on

$A_{j-1} = \mathcal{S}(F_j)$ are also independent. We can see that this is always true, as follows. We have chosen a complement F_j of E_{j-1} in E_j. Suppose we choose another complement F_j' of E_{j-1} in E_j, and let $A_{j-1}' = \mathcal{S}(F_j')$. Then $\mathcal{S}(E_{j-1} \oplus F_j) = \mathcal{S}(E_j) = \mathcal{S}(E_{j-1} \oplus F_j')$. But $\mathcal{S}(E_{j-1}) \cap \mathcal{S}(F_j) = \{0\}$ as every nonzero vector in $\mathcal{S}(E_{j-1})$ has index at most $j - 2$ while every nonzero vector in $\mathcal{S}(F_j)$ has index $j - 1$. Then $\mathcal{S}(E_{j-1}) \cap \mathcal{S}(F_j) = \{0\}$ for exactly the same reason. Thus A_{j-1} and A_{j-1}' are both complements of E_{j-1} in E_j and hence both have the same dimension. But we know there is a Jordan basis, so for the "right" choice of F_j' the labels on the nodes on A_{j-1}' are independent, and then this must be true for our choice F_j as well. ◇

Remark 3.3.5. N_j is a complement of $E_{j-1} \oplus A_j$ in E_j. Note that, in general, a subspace of a vector space will have infinitely many different complements, as there are, in general, infinitely many ways to extend a basis of a subspace to a basis of a vector space. Thus, in general, there can be no unique choice of N_j, and, even if there were, there would be, in general, infinitely many ways to choose a basis for this subspace. Thus we cannot expect any kind of uniqueness result for the ℓESP of $\mathcal{T}$. ◇

3.4 EXAMPLES

In this section, we present a million examples (well, not quite that many) of our algorithm.

Example 3.4.1. Let $A = \begin{bmatrix} 22 & -84 \\ 6 & -23 \end{bmatrix}$, with $c_A(x) = (x + 2)(x - 1)$.

Then A has an eigenvalue -2 of multiplicity 1, and an eigenvalue 1 of multiplicity 1.

We know immediately from Corollary 2.4.3 that A is diagonalizable, so $m_A(x) = (x + 2)(x - 1)$ and A has ℓESP

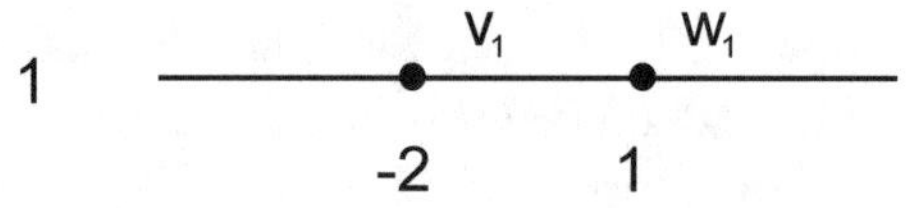

Figure 3.6.

Now we must find the labels. $E(-2) = E_1(-2)$ has basis $\left\{ \begin{bmatrix} 7 \\ 2 \end{bmatrix} \right\}$, and $E(1) = E_1(1)$ has basis $\left\{ \begin{bmatrix} 4 \\ 1 \end{bmatrix} \right\}$. Then we choose $v_1 = \begin{bmatrix} 7 \\ 2 \end{bmatrix}$, and $w_1 = \begin{bmatrix} 4 \\ 1 \end{bmatrix}$.

Then $A = PJP^{-1}$ with

$$P = \begin{bmatrix} 7 & 4 \\ 2 & 1 \end{bmatrix} \text{ and } J = \begin{bmatrix} -2 & 0 \\ 0 & 1 \end{bmatrix}.$$

Example 3.4.2. Let $A = \begin{bmatrix} 7 & 2 \\ -8 & -1 \end{bmatrix}$, with $c_A(x) = (x - 3)^2$.

Then A has an eigenvalue 3 of multiplicity 2.

$E_1(3)$ is 1 dimensional with basis $\left\{ \begin{bmatrix} -2 \\ 1 \end{bmatrix} \right\}$, and $E_2(3)$ is 2 dimensional with basis $\left\{ \begin{bmatrix} 1 \\ 0 \end{bmatrix}, \begin{bmatrix} 0 \\ 1 \end{bmatrix} \right\}$.

Thus $d_1(3) = 1, d_2(3) = 2$, so $d_2^{ex}(3) = 1, d_1^{ex}(3) = 1$, and $d_2^{new}(3) = 1, d_1^{new}(3) = 0$.
Thus $m_A(x) = (x-3)^2$ and A has ℓESP

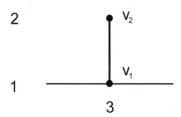

Figure 3.7.

Now we must find the labels. We can choose v_2 to be any vector in $E_2(3)$ that is not in $E_1(3)$.
We choose $v_2 = \begin{bmatrix} 1 \\ 0 \end{bmatrix}$, and then $v_1 = (A - 3I)(v_2) = \begin{bmatrix} 4 \\ -8 \end{bmatrix}$.

Then $A = PJP^{-1}$ with

$$P = \begin{bmatrix} 4 & 1 \\ -8 & 0 \end{bmatrix} \text{ and } J = \begin{bmatrix} 3 & 1 \\ 0 & 3 \end{bmatrix}.$$

Example 3.4.3. Let $A = \begin{bmatrix} 2 & -3 & -3 \\ 2 & -2 & -2 \\ -2 & 1 & 1 \end{bmatrix}$, with $c_A(x) = (x+1)x(x-2)$.

Then A has an eigenvalue -1 of multiplicity 1, an eigenvalue 0 of multiplicity 1, and an eigenvalue 2 of multiplicity 1.

We know immediately from Corollary 2.4.3 that A is diagonalizable, so $m_A(x) = (x + 1)x(x - 2)$ and A has ℓESP $E_1(3)$ is 1 dimensional with basis

$$
\begin{array}{ccc}
 & V_1 & W_1 & X_1 \\
1 & \bullet & \bullet & \bullet \\
 & -1 & 0 & 2
\end{array}
$$

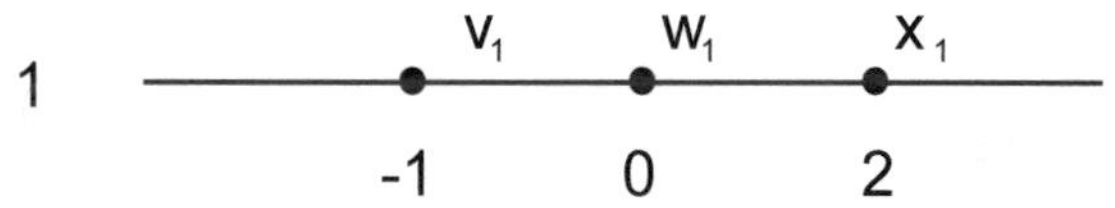

Figure 3.8.

Now we must find the labels. $E(-1)$ has basis $\left\{ \begin{bmatrix} 1 \\ 0 \\ 1 \end{bmatrix} \right\}$, $E(0)$ has basis $\left\{ \begin{bmatrix} 0 \\ 1 \\ -1 \end{bmatrix} \right\}$, and $E(2)$

has basis $\left\{ \begin{bmatrix} 1 \\ 1 \\ -1 \end{bmatrix} \right\}$. Then we choose $v_1 = \begin{bmatrix} 1 \\ 0 \\ 1 \end{bmatrix}$, $w_1 = \begin{bmatrix} 0 \\ 1 \\ -1 \end{bmatrix}$, and $x_1 = \begin{bmatrix} 1 \\ 1 \\ -1 \end{bmatrix}$.

Then $A = PJP^{-1}$ with

$$P = \begin{bmatrix} 1 & 0 & 1 \\ 0 & -1 & 1 \\ 1 & 1 & -1 \end{bmatrix} \text{ and } J = \begin{bmatrix} -1 & 0 & 0 \\ 0 & 0 & 0 \\ 0 & 0 & 2 \end{bmatrix}.$$

Example 3.4.4. Let $A = \begin{bmatrix} 3 & 1 & 1 \\ 2 & 4 & 2 \\ 1 & 1 & 3 \end{bmatrix}$, with $c_A(x) = (x - 2)^2(x - 6)$.

Then A has an eigenvalue 2 of multiplicity 2, and an eigenvalue 6 of multiplicity 1.

$E_1(2)$ is 2 dimensional with basis $\left\{ \begin{bmatrix} 1 \\ -1 \\ 0 \end{bmatrix}, \begin{bmatrix} 0 \\ 1 \\ -1 \end{bmatrix} \right\}$, and $E_1(6)$ is 1 dimensional with basis

$\left\{ \begin{bmatrix} 1 \\ 2 \\ 1 \end{bmatrix} \right\}.$

Thus by Theorem 2.4.2 A is diagonalizable, so $m_A(x) = (x - 2)(x - 6)$ and A has ℓESP

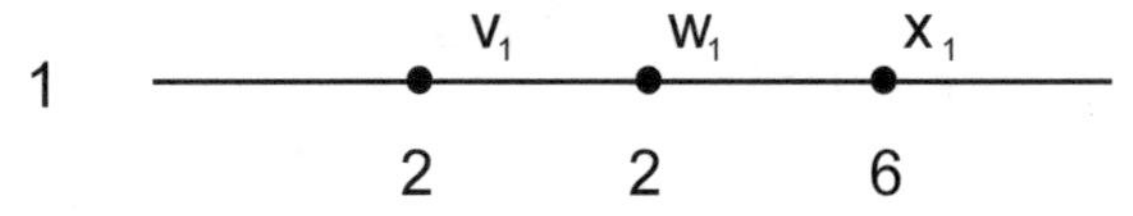

Figure 3.9.

Now we must find the labels. Given the bases we have found above, we choose $v_1 = \begin{bmatrix} 1 \\ -1 \\ 0 \end{bmatrix}$,

$w_1 = \begin{bmatrix} 0 \\ 1 \\ -1 \end{bmatrix}$, and $x_1 = \begin{bmatrix} 1 \\ 2 \\ 1 \end{bmatrix}$.

Then $A = PJP^{-1}$ with

$$P = \begin{bmatrix} 1 & 0 & 1 \\ -1 & 1 & 2 \\ 0 & -1 & 1 \end{bmatrix} \text{ and } J = \begin{bmatrix} 2 & 0 & 0 \\ 0 & 2 & 0 \\ 0 & 0 & 6 \end{bmatrix}.$$

Example 3.4.5. Let $A = \begin{bmatrix} 2 & 1 & 1 \\ 2 & 1 & -2 \\ -1 & 0 & -2 \end{bmatrix}$, with $c_A(x) = (x+1)^2(x-3)$.

Then A has an eigenvalue -1 of multiplicity 2, and an eigenvalue 3 of multiplicity 1.

$E_1(-1)$ is 1 dimensional with basis $\left\{ \begin{bmatrix} 1 \\ -2 \\ 1 \end{bmatrix} \right\}$, and $E_2(-1)$ is 2 dimensional with basis

$\left\{ \begin{bmatrix} 1 \\ -2 \\ 0 \end{bmatrix}, \begin{bmatrix} 0 \\ 0 \\ 1 \end{bmatrix} \right\}$. $E_1(3)$ is 1 dimensional with basis $\left\{ \begin{bmatrix} 5 \\ 6 \\ -1 \end{bmatrix} \right\}$.

Thus $d_1(-1) = 1$, $d_2(-1) = 2$, so $d_2^{ex}(-1) = 1$, $d_1^{ex}(-1) = 1$, and $d_2^{new}(-1) = 1$, $d_1^{new}(-1) = 0$. Also $d_1(3) = d_1^{ex}(3) = d_1^{new}(3) = 1$.
Thus $m_A(x) = (x+1)^2(x-3)$ and A has ℓESP

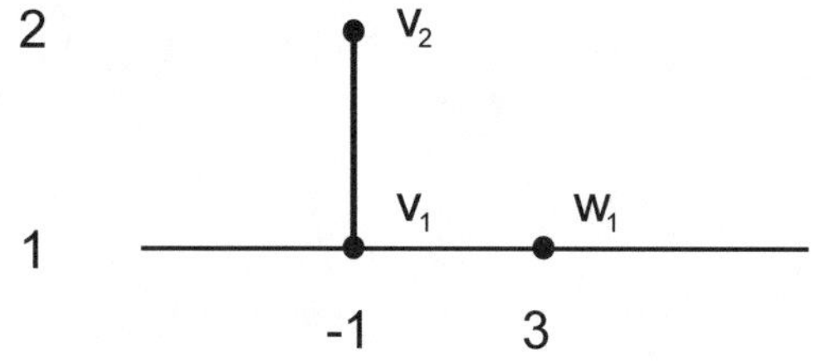

Figure 3.10.

Now we must find the labels. We must choose v_2 to be a vector that is in $E_2(-1)$ but that is not in $E_1(-1)$. Given the bases we have found above, we choose $v_2 = \begin{bmatrix} 1 \\ -2 \\ 0 \end{bmatrix}$, and then

$v_1 = (A - (-1)I)v_2 = \begin{bmatrix} 1 \\ -2 \\ 1 \end{bmatrix}$. (It is purely a coincidence that v_1 was a vector in the basis we chose for $E_1(-1)$, and that will in general not be the case. Compare Example 3.4.2, where it wasn't.) Also, we choose $w_1 = \begin{bmatrix} 5 \\ 6 \\ -1 \end{bmatrix}$.

Then $A = PJP^{-1}$ with

$$P = \begin{bmatrix} 1 & 1 & 5 \\ -2 & -2 & 6 \\ 1 & 0 & -1 \end{bmatrix} \text{ and } J = \begin{bmatrix} -1 & 1 & 0 \\ 0 & -1 & 0 \\ 0 & 0 & 3 \end{bmatrix}.$$

Example 3.4.6. Let $A = \begin{bmatrix} 2 & 1 & 1 \\ -2 & -1 & -2 \\ 1 & 1 & 2 \end{bmatrix}$, with $c_A(x) = (x-1)^3$.

Then A has an eigenvalue 1 of multiplicity 3.

$E_1(1)$ is 2 dimensional with basis $\left\{ \begin{bmatrix} 1 \\ 0 \\ -1 \end{bmatrix}, \begin{bmatrix} 0 \\ 1 \\ -1 \end{bmatrix} \right\}$. and $E_2(1)$ is 3 dimensional with basis

$\left\{ \begin{bmatrix} 1 \\ 0 \\ 0 \end{bmatrix}, \begin{bmatrix} 0 \\ 1 \\ 0 \end{bmatrix}, \begin{bmatrix} 0 \\ 0 \\ 1 \end{bmatrix} \right\}$.

Thus $d_1(1) = 2, d_2(1) = 3$, so $d_2^{ex}(1) = 1, d_1^{ex}(1) = 2$, and $d_2^{new}(1) = 1, d_1^{new}(1) = 1$. Thus $m_A(x) = (x-1)^2$ and A has ℓESP

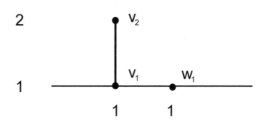

Figure 3.11.

Now we must find the labels. We must choose v_2 to be a vector that is in $E_2(1)$ but that is not in $E_1(1)$. Given the bases we have found above, we choose $v_2 = \begin{bmatrix} 1 \\ 0 \\ 0 \end{bmatrix}$, and then $v_1 = (A - (-1)I)v_2 =$

$\begin{bmatrix} 1 \\ -2 \\ 1 \end{bmatrix}$. (Since v_1 is in $E_1(1)$, it must be a linear combination of our basis elements, and, sure enough,

$\begin{bmatrix} 1 \\ -2 \\ 1 \end{bmatrix} = (-1) \begin{bmatrix} 1 \\ 0 \\ 0 \end{bmatrix} + (-2) \begin{bmatrix} 1 \\ 0 \\ 0 \end{bmatrix}$.) Also, we may choose w_1 to be any vector in $E_1(1)$ such that

$\{v_1, w_1\}$ is linearly independent, and we choose $w_1 = \begin{bmatrix} 1 \\ 0 \\ -1 \end{bmatrix}$.

Then $A = PJP^{-1}$ with

$$P = \begin{bmatrix} 1 & 1 & 1 \\ -2 & 0 & 0 \\ 1 & 0 & -1 \end{bmatrix} \text{ and } J = \begin{bmatrix} 1 & 1 & 0 \\ 0 & 1 & 0 \\ 0 & 0 & 1 \end{bmatrix}.$$

Example 3.4.7. Let $A = \begin{bmatrix} 5 & 0 & 1 \\ 1 & 1 & 0 \\ -7 & 1 & 0 \end{bmatrix}$, with $c_A(x) = (x-2)^3$.

Then A has an eigenvalue 2 of multiplicity 3.

$E_1(2)$ is 1 dimensional with basis $\left\{ \begin{bmatrix} 1 \\ 1 \\ -3 \end{bmatrix} \right\}$, $E_2(2)$ is 2 dimensional with basis

$\left\{ \begin{bmatrix} 1 \\ 0 \\ -2 \end{bmatrix}, \begin{bmatrix} 0 \\ 1 \\ -1 \end{bmatrix} \right\}$, and $E_3(2)$ is 3 dimensional with basis $\left\{ \begin{bmatrix} 1 \\ 0 \\ 0 \end{bmatrix}, \begin{bmatrix} 0 \\ 1 \\ 0 \end{bmatrix}, \begin{bmatrix} 0 \\ 0 \\ 1 \end{bmatrix} \right\}$.

Thus $d_1(2) = 1, d_2(2) = 2, d_3(2) = 3$, so $d_3^{\text{ex}}(2) = 1, d_2^{\text{ex}}(2) = 1, d_1^{\text{ex}}(2) = 2$, and $d_3^{\text{new}}(2) = 1, d_2^{\text{new}}(2) = 0, d_1^{\text{new}}(2) = 0$.

Thus $m_A(x) = (x-2)^3$ and A has ℓESP

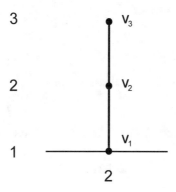

Figure 3.12.

Now we must find the labels. We must choose v_3 to be a vector that is in $E_2(3)$ but that is not in $E_2(3)$. Given the bases we have found above, we choose $v_3 = \begin{bmatrix} 1 \\ 0 \\ 0 \end{bmatrix}$, then $v_2 = (A - 2I)v_2 = \begin{bmatrix} 3 \\ 1 \\ -7 \end{bmatrix}$, and $v_1 = (A - 2)I)v_2 = \begin{bmatrix} 2 \\ 2 \\ -6 \end{bmatrix}$.

Then $A = PJP^{-1}$ with

$$P = \begin{bmatrix} 2 & 3 & 1 \\ 2 & 1 & 0 \\ -6 & -7 & 0 \end{bmatrix} \text{ and } J = \begin{bmatrix} 2 & 1 & 0 \\ 0 & 2 & 1 \\ 0 & 0 & 2 \end{bmatrix}.$$

Example 3.4.8. Let $A = \begin{bmatrix} 0 & 1 & 1 & -1 \\ 12 & 6 & -6 & 6 \\ 6 & 2 & -1 & 3 \\ 16 & 0 & -8 & 10 \end{bmatrix}$, with $c_A(x) = (x - 2)^3(x - 5)$.

Then A has an eigenvalue 2 of multiplicity 3, and an eigenvalue 5 of multiplicity 1.

$E_1(2)$ is 2 dimensional with basis $\left\{ \begin{bmatrix} 1 \\ 0 \\ 2 \\ 0 \end{bmatrix}, \begin{bmatrix} 0 \\ 0 \\ 1 \\ 1 \end{bmatrix} \right\}$, and $E_2(2)$ is 3 dimensional with basis

$\left\{ \begin{bmatrix} 1 \\ 0 \\ 2 \\ 0 \end{bmatrix}, \begin{bmatrix} 0 \\ 1 \\ 0 \\ 0 \end{bmatrix}, \begin{bmatrix} 0 \\ 0 \\ 1 \\ 1 \end{bmatrix} \right\}$. $E_1(5)$ is 1 dimensional with basis $\left\{ \begin{bmatrix} 1 \\ 6 \\ 7 \\ 8 \end{bmatrix} \right\}$,

Thus $d_1(2) = 2$, $d_2(2) = 3$, so $d_2^{ex}(2) = 1$, $d_1^{ex}(2) = 2$, and $d_2^{new}(2) = 1$, $d_1^{new}(2) = 1$. Also, $d_1(5) = d_1^{ex}(5) = d_1^{new}(2) = 5$.

Thus $m_A(x) = (x - 1)^2(x - 5)$ and A has ℓESP

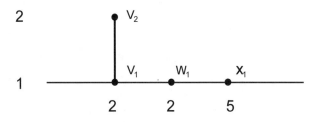

Figure 3.13.

Now we must find the labels. We must choose v_2 to be a vector that is in $E_2(2)$ but that is not

in $E_1(2)$. Given the bases we have found above, we choose $v_2 = \begin{bmatrix} 0 \\ 1 \\ 0 \\ 0 \end{bmatrix}$, and then $v_1 = (A - 2I)v_2 =$

$\begin{bmatrix} 1 \\ 0 \\ 2 \\ 0 \end{bmatrix}$. We must choose w_1 to be a vector that is in $E_1(2)$ with $\{v_1, w_1\}$ linearly independent, and we

choose $w_1 = \begin{bmatrix} 0 \\ 0 \\ 1 \\ 1 \end{bmatrix}$. We choose $x_1 = \begin{bmatrix} 1 \\ 6 \\ 7 \\ 8 \end{bmatrix}$.

Then $A = PJP^{-1}$ with

$$P = \begin{bmatrix} 1 & 0 & 0 & 1 \\ 0 & 1 & 0 & 6 \\ 2 & 0 & 1 & 7 \\ 0 & 0 & 1 & 8 \end{bmatrix} \text{ and } J = \begin{bmatrix} 2 & 1 & 0 & 0 \\ 0 & 2 & 0 & 0 \\ 0 & 0 & 2 & 0 \\ 0 & 0 & 0 & 5 \end{bmatrix}.$$

Example 3.4.9. Let $A = \begin{bmatrix} -1 & 1 & 0 & 1 \\ 0 & 3 & 1 & 0 \\ -6 & -4 & -1 & 3 \\ -6 & 2 & 0 & 4 \end{bmatrix}$, with $c_A(x) = (x - 1)^3(x - 2)$.

Then A has an eigenvalue 1 of multiplicity 3, and an eigenvalue 2 of multiplicity 1.

$E_1(1)$ is 1 dimensional with basis $\left\{ \begin{bmatrix} 1 \\ 0 \\ 0 \\ 2 \end{bmatrix} \right\}$, $E_2(1)$ is 2 dimensional with basis

$\left\{ \begin{bmatrix} 1 \\ 0 \\ 0 \\ 2 \end{bmatrix}, \begin{bmatrix} 0 \\ 1 \\ -2 \\ 0 \end{bmatrix} \right\}$, and $E_3(1)$ is 3 dimensional with basis $\left\{ \begin{bmatrix} 1 \\ 0 \\ 0 \\ 2 \end{bmatrix}, \begin{bmatrix} 0 \\ 1 \\ 0 \\ 0 \end{bmatrix}, \begin{bmatrix} 0 \\ 0 \\ 1 \\ 0 \end{bmatrix} \right\}$. $E_1(2)$ is 1 dimen-

sional with basis $\left\{ \begin{bmatrix} 4 \\ 3 \\ -3 \\ 9 \end{bmatrix} \right\}$.

Thus $d_1(1) = 1, d_2(1) = 2, d_3(1) = 3$, so $d_3^{ex}(1) = 1, d_2^{ex}(1) = 1, d_1^{ex}(1) = 1$, and $d_3^{new}(1) =$
$1, d_2^{new}(1) = 0, d_1^{new}(1) = 0$. Also, $d_1(1) = d_1^{ex}(1) = d_1^{new}(1) = 5$.
Thus $m_A(x) = (x - 1)^3(x - 2)$ and A has ℓESP

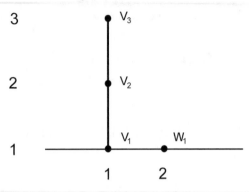

Figure 3.14.

Now we must find the labels. We must choose v_3 to be a vector that is in $E_3(1)$ but that is not in $E_2(1)$.

Given the bases we have found above, we choose $v_3 = \begin{bmatrix} 0 \\ 1 \\ 0 \\ 0 \end{bmatrix}$, then $v_2 = (A - 1I)v_3 = \begin{bmatrix} 1 \\ 2 \\ -4 \\ 2 \end{bmatrix}$, and

then $v_1 = (A - 1I)v_2 = \begin{bmatrix} 2 \\ 0 \\ 0 \\ 4 \end{bmatrix}$. Also, we choose $w_1 = \begin{bmatrix} 4 \\ 3 \\ -3 \\ 9 \end{bmatrix}$.

Then $A = PJP^{-1}$ with

$$P = \begin{bmatrix} 2 & 1 & 0 & 4 \\ 0 & 2 & 1 & 3 \\ 0 & -4 & 0 & -3 \\ 4 & 2 & 0 & 9 \end{bmatrix} \text{ and } J = \begin{bmatrix} 1 & 1 & 0 & 0 \\ 0 & 1 & 1 & 0 \\ 0 & 0 & 1 & 0 \\ 0 & 0 & 0 & 2 \end{bmatrix}.$$

Example 3.4.10. Let $A = \begin{bmatrix} 4 & 1 & -1 & 2 \\ -2 & 7 & -2 & 4 \\ 3 & -3 & 8 & -6 \\ 2 & -2 & 2 & 1 \end{bmatrix}$, with $c_A(x) = (x - 5)^4$.

Then A has an eigenvalue 5 of multiplicity 4.

$E_1(5)$ is 3 dimensional with basis $\left\{ \begin{bmatrix} 2 \\ 0 \\ 0 \\ 1 \end{bmatrix}, \begin{bmatrix} 0 \\ 2 \\ 0 \\ -1 \end{bmatrix}, \begin{bmatrix} 0 \\ 0 \\ 2 \\ 1 \end{bmatrix} \right\}$, and $E_2(5)$ is 4 dimensional with

basis $\left\{ \begin{bmatrix} 1 \\ 0 \\ 0 \\ 0 \end{bmatrix}, \begin{bmatrix} 0 \\ 1 \\ 0 \\ 0 \end{bmatrix}, \begin{bmatrix} 0 \\ 0 \\ 1 \\ 0 \end{bmatrix}, \begin{bmatrix} 0 \\ 0 \\ 0 \\ 1 \end{bmatrix} \right\}$.

Thus $d_1(5) = 3$, $d_2(5) = 4$, so $d_2^{ex}(5) = 1$, $d_1^{ex}(5) = 3$, and $d_2^{new}(5) = 1$, $d_1^{new}(5) = 2$.
Thus $m_A(x) = (x - 5)^2$ and A has ℓESP

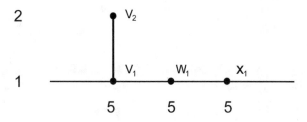

Figure 3.15.

Now we must find the labels. We must choose v_2 to be a vector that is in $E_2(5)$ but that is not

in $E_1(5)$. Given the bases we have found above, we choose $v_2 = \begin{bmatrix} 1 \\ 0 \\ 0 \\ 0 \end{bmatrix}$, and then $v_1 = (A - 5I)v_2 =$

$\begin{bmatrix} -1 \\ -2 \\ 3 \\ 2 \end{bmatrix}$. Then we must choose vectors w_1 and x_1 in $E_1(5)$ so that $\{v_1, w_1, x_1\}$ is linearly independent,

and we choose $w_1 = \begin{bmatrix} 2 \\ 0 \\ 0 \\ 1 \end{bmatrix}$ and $x_1 = \begin{bmatrix} 0 \\ 2 \\ 0 \\ -1 \end{bmatrix}$.

Then $A = PJP^{-1}$ with

$$P = \begin{bmatrix} -1 & 1 & 2 & 0 \\ -2 & 0 & 0 & 2 \\ 3 & 0 & 0 & 0 \\ 2 & 0 & 1 & -1 \end{bmatrix} \text{ and } J = \begin{bmatrix} 5 & 1 & 0 & 0 \\ 0 & 5 & 0 & 0 \\ 0 & 0 & 5 & 0 \\ 0 & 0 & 0 & 5 \end{bmatrix}.$$

Example 3.4.11. Let $A = \begin{bmatrix} 7 & -2 & -1 & 3 \\ 1 & 3 & 0 & 1 \\ 1 & -1 & 4 & 1 \\ -2 & 1 & 1 & 2 \end{bmatrix}$, with $c_A(x) = (x-4)^4$.

Then A has an eigenvalue 4 of multiplicity 4.

$E_1(4)$ is 2 dimensional with basis $\left\{ \begin{bmatrix} 1 \\ 1 \\ 1 \\ 0 \end{bmatrix}, \begin{bmatrix} 0 \\ 1 \\ 1 \\ 1 \end{bmatrix} \right\} = \{t_1, u_1\}$, and $E_2(4)$ is 4 dimensional with

basis $\left\{ \begin{bmatrix} 1 \\ 0 \\ 0 \\ 0 \end{bmatrix}, \begin{bmatrix} 0 \\ 1 \\ 0 \\ 0 \end{bmatrix}, \begin{bmatrix} 0 \\ 0 \\ 1 \\ 0 \end{bmatrix}, \begin{bmatrix} 0 \\ 0 \\ 0 \\ 1 \end{bmatrix} \right\}.$

Thus $d_1(4) = 2$, $d_2(4) = 4$, so $d_2^{ex}(4) = 2$, $d_1^{ex}(4) = 2$, and $d_2^{new}(4) = 2$, $d_1^{new}(4) = 0$.
Thus $m_A(x) = (x-4)^4$ and A has ℓESP

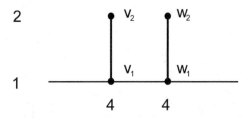

Figure 3.16.

Now we must find the labels. We must choose v_2 and w_2 to be vectors in $E_2(4)$ with

$\{v_2, w_2, t_1, u_1\}$ linearly independent. Given the bases we have found above, we choose $v_2 = \begin{bmatrix} 1 \\ 0 \\ 0 \\ 0 \end{bmatrix}$

and $w_2 = \begin{bmatrix} 0 \\ 1 \\ 0 \\ 0 \end{bmatrix}$. Then $v_1 = (A - 4I)v_2 = \begin{bmatrix} 3 \\ 1 \\ 1 \\ -2 \end{bmatrix}$ and $w_1 = \begin{bmatrix} -2 \\ -1 \\ -1 \\ 1 \end{bmatrix}$.

Then $A = PJP^{-1}$ with

$$P = \begin{bmatrix} 3 & 1 & -2 & 0 \\ 1 & 0 & -1 & 1 \\ 1 & 0 & -1 & 0 \\ -2 & 0 & 1 & 0 \end{bmatrix} \text{ and } J = \begin{bmatrix} 4 & 1 & 0 & 0 \\ 0 & 4 & 0 & 0 \\ 0 & 0 & 4 & 1 \\ 0 & 0 & 0 & 4 \end{bmatrix}.$$

Example 3.4.12. Let $A = \begin{bmatrix} -2 & -7 & -10 & 5 & 8 & 5 \\ -1 & 1 & -1 & 1 & 2 & 1 \\ -1 & -3 & 1 & 1 & 3 & 1 \\ -4 & -7 & -8 & 7 & 7 & 4 \\ -2 & -1 & -3 & 2 & 4 & 2 \\ -2 & -6 & -4 & 2 & 7 & 5 \end{bmatrix}$, with $c_A(x) = (x-3)^4(x-2)^2$.

Then A has an eigenvalue 3 of multiplicity 4, and an eigenvalue 2 of multiplicity 2.

$E_1(3)$ is 2 dimensional with basis $\left\{ \begin{bmatrix} 1 \\ 0 \\ 0 \\ 0 \\ 0 \\ 1 \end{bmatrix}, \begin{bmatrix} 0 \\ 0 \\ 0 \\ 1 \\ 0 \\ -1 \end{bmatrix} \right\}$, $E_2(3)$ is 3 dimensional with basis

$\left\{ \begin{bmatrix} 1 \\ 0 \\ 0 \\ 0 \\ 0 \\ 1 \end{bmatrix}, \begin{bmatrix} 0 \\ 1 \\ 0 \\ 0 \\ 1 \\ 0 \end{bmatrix}, \begin{bmatrix} 0 \\ 0 \\ 0 \\ 1 \\ 0 \\ -1 \end{bmatrix} \right\}$, and $E_3(3)$ is 4 dimensional with basis $\left\{ \begin{bmatrix} 1 \\ 0 \\ 0 \\ 0 \\ 0 \\ 1 \end{bmatrix}, \begin{bmatrix} 0 \\ 1 \\ 0 \\ 0 \\ 1 \\ 0 \end{bmatrix}, \begin{bmatrix} 0 \\ 0 \\ 1 \\ 0 \\ 0 \\ 2 \end{bmatrix}, \begin{bmatrix} 0 \\ 0 \\ 0 \\ 1 \\ 0 \\ -1 \end{bmatrix} \right\}$.

Also, $E_1(2)$ is 1 dimensional with basis $\left\{ \begin{bmatrix} 9 \\ -1 \\ 3 \\ 7 \\ -2 \\ 8 \end{bmatrix} \right\}$, and $E_2(2)$ is 2 dimensional with basis

$\left\{ \begin{bmatrix} -18 \\ 0 \\ -8 \\ -17 \\ 4 \\ -18 \end{bmatrix}, \begin{bmatrix} 0 \\ 2 \\ 2 \\ 3 \\ 0 \\ 2 \end{bmatrix} \right\}$.

Thus $d_1(3) = 2, d_2(3) = 3, d_3(3) = 4$, so $d_3^{ex}(3) = 1, d_2^{ex}(3) = 1, d_1^{ex}(3) = 2$, and $d_3^{new}(3) = 1, d_2^{new}(3) = 0, d_1^{new}(3) = 1$. Also, $d_1(2) = 1, d_2(2) = 2$, so $d_2^{ex}(2) = 1, d_1^{ex}(2) = 1$, and $d_2^{new}(2) = 1, d_1^{new}(2) = 0$.

Thus $m_A(x) = (x-3)^3(x-2)^2$ and A has ℓESP

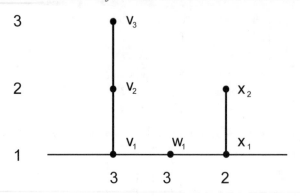

Figure 3.17.

Now we must find the labels. We must choose v_3 to be a vector that is in $E_3(3)$ but that is not in $E_2(3)$. Given the bases we have found above, we choose $v_3 = \begin{bmatrix} 0 \\ 0 \\ 1 \\ 0 \\ 0 \\ 2 \end{bmatrix}$ and then $v_2 = (A - 3I)v_3 =$

$\begin{bmatrix} 0 \\ 1 \\ 0 \\ 0 \\ 1 \\ 0 \end{bmatrix}$ and $v_1 = (A - 3I)v_2 = \begin{bmatrix} 1 \\ 0 \\ 0 \\ 0 \\ 0 \\ 1 \end{bmatrix}$. Then we must choose w_1 to be a vector in $E_1(3)$ such that

$\{v_1, w_1\}$ is linearly independent, and we choose $w_1 = \begin{bmatrix} 0 \\ 0 \\ 0 \\ 1 \\ 0 \\ -1 \end{bmatrix}$. Also, we must choose x_2 to be a vector

in $E_2(2)$ that is not in $E_1(2)$. We choose $x_2 = \begin{bmatrix} -18 \\ 0 \\ -8 \\ -17 \\ 4 \\ -18 \end{bmatrix}$, and then $x_1 = (A - 2I)x_2 = \begin{bmatrix} 9 \\ -1 \\ 3 \\ 7 \\ -2 \\ 8 \end{bmatrix}$.

Then $A = PJP^{-1}$ with

$$
P = \begin{bmatrix}
1 & 0 & 0 & 0 & 9 & -18 \\
0 & 1 & 0 & 0 & -1 & 0 \\
0 & 0 & 1 & 0 & 3 & -8 \\
0 & 0 & 0 & 1 & 7 & -17 \\
0 & 1 & 0 & 0 & -2 & 4 \\
1 & 0 & 2 & -1 & 8 & -18
\end{bmatrix}
\quad \text{and } J = \begin{bmatrix}
3 & 1 & 0 & 0 & 0 & 0 \\
0 & 3 & 1 & 0 & 0 & 0 \\
0 & 0 & 3 & 0 & 0 & 0 \\
0 & 0 & 0 & 3 & 0 & 0 \\
0 & 0 & 0 & 0 & 2 & 1 \\
0 & 0 & 0 & 0 & 0 & 2
\end{bmatrix}.
$$

Example 3.4.13. Let $A = \begin{bmatrix}
13 & -3 & 1 & -3 & 9 & 3 \\
3 & 2 & 1 & -1 & -3 & 1 \\
9 & -4 & 5 & -3 & -9 & 3 \\
15 & -7 & 2 & -1 & -15 & 5 \\
18 & -5 & 0 & -6 & -14 & 6 \\
42 & -14 & 0 & -14 & -42 & 18
\end{bmatrix}$, with $c_A(x) = (x-4)^5(x-3)$.

Then A has an eigenvalue 4 of multiplicity 5, and an eigenvalue 3 of multiplicity 1.

$E_1(4)$ is 3 dimensional with basis $\left\{ \begin{bmatrix} 1 \\ 0 \\ 0 \\ 0 \\ 0 \\ -3 \end{bmatrix}, \begin{bmatrix} 0 \\ 0 \\ 0 \\ 1 \\ 0 \\ 1 \end{bmatrix}, \begin{bmatrix} 0 \\ 0 \\ 0 \\ 0 \\ 1 \\ 3 \end{bmatrix} \right\}$, $E_2(4)$ is 4 dimen-

sional with basis $\left\{ \begin{bmatrix} 1 \\ 0 \\ 0 \\ 0 \\ 0 \\ -3 \end{bmatrix}, \begin{bmatrix} 0 \\ 1 \\ 1 \\ 0 \\ 0 \\ 1 \end{bmatrix}, \begin{bmatrix} 0 \\ 0 \\ 0 \\ 1 \\ 0 \\ 1 \end{bmatrix}, \begin{bmatrix} 0 \\ 0 \\ 0 \\ 0 \\ 1 \\ 3 \end{bmatrix} \right\}$, and $E_3(4)$ is 5 dimensional with basis

$\left\{ \begin{bmatrix} 1 \\ 0 \\ 0 \\ 0 \\ 0 \\ -3 \end{bmatrix}, \begin{bmatrix} 0 \\ 1 \\ 0 \\ 0 \\ 0 \\ 1 \end{bmatrix}, \begin{bmatrix} 0 \\ 0 \\ 1 \\ 0 \\ 0 \\ 0 \end{bmatrix}, \begin{bmatrix} 0 \\ 0 \\ 0 \\ 1 \\ 0 \\ 1 \end{bmatrix}, \begin{bmatrix} 0 \\ 0 \\ 0 \\ 0 \\ 1 \\ 3 \end{bmatrix} \right\}$. Also, $E_1(3)$ is 1 dimensional with basis $\left\{ \begin{bmatrix} 2 \\ -1 \\ 1 \\ 1 \\ 7 \\ 14 \end{bmatrix} \right\}$.

Thus $d_1(4) = 3, d_2(4) = 4, d_3(4) = 5$, so $d_3^{\text{ex}}(4) = 1, d_2^{\text{ex}}(4) = 1, d_1^{\text{ex}}(4) = 3$, and $d_3^{\text{new}}(4) = 1, d_2^{\text{new}}(4) = 0, d_1^{\text{new}}(4) = 2$. Also, $d_1(3) = d_1^{\text{ex}}(3) = d_1^{\text{new}}(3) = 1$.

Thus $m_A(x) = (x-4)^3(x-3)$ and A has ℓESP

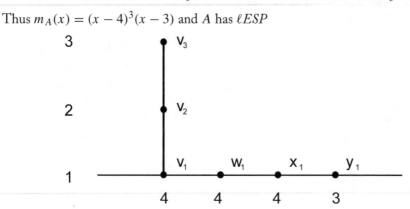

Figure 3.18.

Now we must find the labels. We must choose v_3 to be a vector that is in $E_3(4)$ but that is not

in $E_2(4)$. Given the bases we have found above, we choose $v_3 = \begin{bmatrix} 0 \\ 1 \\ 0 \\ 0 \\ 0 \\ 1 \end{bmatrix}$ and then $v_2 = (A-4I)v_3 =$

$\begin{bmatrix} 0 \\ -1 \\ -1 \\ -2 \\ 1 \\ 0 \end{bmatrix}$ and $v_1 = (A-4I)v_2 = \begin{bmatrix} -1 \\ 0 \\ 0 \\ 0 \\ -1 \\ 0 \end{bmatrix}$. Then we must choose w_1 and x_1 to be vectors in $E_1(4)$ such

that $\{v_1, w_1, x_1\}$ is linearly independent, and we choose $w_1 = \begin{bmatrix} 1 \\ 0 \\ 0 \\ 0 \\ 0 \\ -3 \end{bmatrix}$ and $x_1 = \begin{bmatrix} 0 \\ 0 \\ 0 \\ 1 \\ 0 \\ 1 \end{bmatrix}$. Also, we may

choose y_1 to be any nonzero vector in $E_1(3)$. We choose $y_1 = \begin{bmatrix} 2 \\ -1 \\ 1 \\ 1 \\ 7 \\ 14 \end{bmatrix}$.

Then $A = PJP^{-1}$ with

$$P = \begin{bmatrix} -1 & 0 & 0 & 1 & 0 & 2 \\ 0 & -1 & 1 & 0 & 0 & -1 \\ 0 & -1 & 0 & 0 & 0 & 1 \\ 0 & -2 & 0 & 0 & 1 & 1 \\ -1 & 1 & 0 & 0 & 0 & 7 \\ 0 & 0 & 1 & -3 & 1 & 14 \end{bmatrix} \text{ and } J = \begin{bmatrix} 4 & 1 & 0 & 0 & 0 & 0 \\ 0 & 4 & 1 & 0 & 0 & 0 \\ 0 & 0 & 4 & 0 & 0 & 0 \\ 0 & 0 & 0 & 4 & 0 & 0 \\ 0 & 0 & 0 & 0 & 4 & 0 \\ 0 & 0 & 0 & 0 & 0 & 3 \end{bmatrix}.$$

Example 3.4.14. Let $A = \begin{bmatrix} 1 & 1 & 2 & -1 & 1 & 1 \\ -6 & 3 & 7 & -3 & 0 & 3 \\ -2 & 0 & 5 & -1 & 0 & 1 \\ -10 & 0 & 10 & -2 & 1 & 5 \\ 2 & 0 & -2 & 1 & 3 & -1 \\ -6 & 2 & 6 & -3 & 3 & 6 \end{bmatrix}$, with $c_A(x) = (x-3)^5(x-1)$.

Then A has an eigenvalue 3 of multiplicity 5, and an eigenvalue 1 of multiplicity 1.

$E_1(3)$ is 2 dimensional with basis $\left\{ \begin{bmatrix} 1 \\ 0 \\ 0 \\ 0 \\ 0 \\ 2 \end{bmatrix}, \begin{bmatrix} 0 \\ 0 \\ 0 \\ 1 \\ 0 \\ 1 \end{bmatrix} \right\} = \{t_1, u_1\}$, $E_2(3)$ is 4 dimen-

sional with basis $\left\{ \begin{bmatrix} 1 \\ 0 \\ 0 \\ 0 \\ 0 \\ 2 \end{bmatrix}, \begin{bmatrix} 0 \\ 1 \\ 0 \\ 0 \\ 0 \\ 0 \end{bmatrix}, \begin{bmatrix} 0 \\ 0 \\ 1 \\ 0 \\ 0 \\ 1 \end{bmatrix}, \begin{bmatrix} 0 \\ 0 \\ 0 \\ 1 \\ 1 \\ 0 \end{bmatrix} \right\}$, and $E_3(3)$ is 5 dimensional with basis

$\left\{ \begin{bmatrix} 1 \\ 0 \\ 0 \\ 0 \\ 0 \\ 2 \end{bmatrix}, \begin{bmatrix} 0 \\ 1 \\ 0 \\ 0 \\ 0 \\ 0 \end{bmatrix}, \begin{bmatrix} 0 \\ 0 \\ 1 \\ 0 \\ 0 \\ -2 \end{bmatrix}, \begin{bmatrix} 0 \\ 0 \\ 0 \\ 1 \\ 0 \\ 1 \end{bmatrix}, \begin{bmatrix} 0 \\ 0 \\ 0 \\ 0 \\ 1 \\ 0 \end{bmatrix} \right\}$. Also, $E_1(1)$ is 1 dimensional with basis $\left\{ \begin{bmatrix} 1 \\ 10 \\ 4 \\ 22 \\ -4 \\ 8 \end{bmatrix} \right\}$.

Thus $d_1(3) = 2, d_2(3) = 4, d_3(3) = 5$, so $d_3^{\text{ex}}(3) = 1, d_2^{\text{ex}}(3) = 2, d_1^{\text{ex}}(3) = 2$, and $d_3^{\text{new}}(4) = 1, d_2^{\text{new}}(4) = 2, d_1^{\text{new}}(4) = 0$. Also, $d_1(1) = d_1^{\text{ex}}(1) = d_1^{\text{new}}(1) = 1$.

Thus $m_A(x) = (x - 3)^3(x - 1)$ and A has ℓESP

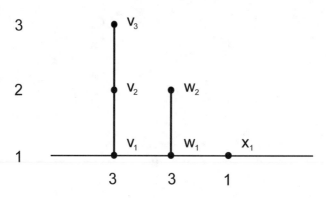

Figure 3.19.

Now we must find the labels. We must choose v_3 to be a vector that is in $E_3(4)$ but that is not

in $E_2(4)$. Given the bases we have found above, we choose $v_3 = \begin{bmatrix} 0 \\ 1 \\ 0 \\ 0 \\ 0 \\ -2 \end{bmatrix}$ and then $v_2 = (A - 4I)v_3 =$

$\begin{bmatrix} 0 \\ 1 \\ 0 \\ 0 \\ 0 \\ 0 \end{bmatrix}$. Now we must choose w_2 in $E_2(3)$ so that $\{v_2, w_2, t_1, u_1\}$ is linearly independent. We choose

$w_2 = \begin{bmatrix} 0 \\ 0 \\ 0 \\ 0 \\ 1 \\ 0 \end{bmatrix}$. Then $v_1 = (A - 3I)v_2 = \begin{bmatrix} 1 \\ 0 \\ 0 \\ 0 \\ 0 \\ 2 \end{bmatrix}$ and $w_1 = (A - 3I)w_2 = \begin{bmatrix} 1 \\ 0 \\ 0 \\ 1 \\ 0 \\ 3 \end{bmatrix}$. Also, we may choose

x_1 to be any nonzero vector in $E_1(3)$. We choose $x_1 = \begin{bmatrix} 1 \\ 10 \\ 4 \\ 22 \\ -4 \\ 8 \end{bmatrix}$.

Then $A = PJP^{-1}$ with

$$P = \begin{bmatrix} 1 & 0 & 0 & 1 & 0 & 1 \\ 0 & 1 & 0 & 0 & 0 & 10 \\ 0 & 0 & 1 & 0 & 0 & 4 \\ 0 & 0 & 0 & 1 & 0 & 22 \\ 0 & 0 & 0 & 0 & 1 & -4 \\ 2 & 0 & -2 & 3 & 0 & 8 \end{bmatrix} \text{ and } J = \begin{bmatrix} 3 & 1 & 0 & 0 & 0 & 0 \\ 0 & 3 & 1 & 0 & 0 & 0 \\ 0 & 0 & 3 & 0 & 0 & 0 \\ 0 & 0 & 0 & 3 & 1 & 0 \\ 0 & 0 & 0 & 0 & 3 & 0 \\ 0 & 0 & 0 & 0 & 0 & 1 \end{bmatrix}.$$

Example 3.4.15. Let $A = \begin{bmatrix} 2 & 3 & 0 & -1 & 2 & -2 \\ -1 & 0 & 2 & 1 & -1 & -2 \\ 1 & 3 & 2 & 0 & 1 & -4 \\ 5 & 6 & -1 & -3 & 5 & 3 \\ 3 & 3 & -1 & -2 & 3 & -1 \\ 1 & 3 & 2 & 0 & 1 & -4 \end{bmatrix}$, with $c_A(x) = x^6$.

Then A has an eigenvalue 0 of multiplicity 6.

$E_1(0)$ is 3 dimensional with basis $\left\{ \begin{bmatrix} 1 \\ 0 \\ 0 \\ 0 \\ -1 \\ 0 \end{bmatrix}, \begin{bmatrix} 0 \\ 1 \\ 0 \\ -3 \\ -3 \\ 0 \end{bmatrix}, \begin{bmatrix} 0 \\ 0 \\ 1 \\ 2 \\ 2 \\ 1 \end{bmatrix} \right\} = \{s_1, t_1, u_1\}$, $E_2(0)$ is 5

dimensional with basis $\left\{ \begin{bmatrix} 1 \\ 0 \\ 0 \\ 0 \\ 0 \\ 0 \end{bmatrix}, \begin{bmatrix} 0 \\ 1 \\ 0 \\ 0 \\ 0 \\ 0 \end{bmatrix}, \begin{bmatrix} 0 \\ 0 \\ 1 \\ 0 \\ 0 \\ 1 \end{bmatrix}, \begin{bmatrix} 0 \\ 0 \\ 0 \\ 1 \\ 0 \\ 0 \end{bmatrix}, \begin{bmatrix} 0 \\ 0 \\ 0 \\ 0 \\ 1 \\ 0 \end{bmatrix} \right\}$, and $E_3(0)$ is 6 dimensional with basis

$$\left\{ \begin{bmatrix} 1 \\ 0 \\ 0 \\ 0 \\ 0 \\ 0 \end{bmatrix}, \begin{bmatrix} 0 \\ 1 \\ 0 \\ 0 \\ 0 \\ 0 \end{bmatrix}, \begin{bmatrix} 0 \\ 0 \\ 1 \\ 0 \\ 0 \\ 0 \end{bmatrix}, \begin{bmatrix} 0 \\ 0 \\ 0 \\ 1 \\ 0 \\ 0 \end{bmatrix}, \begin{bmatrix} 0 \\ 0 \\ 0 \\ 0 \\ 1 \\ 0 \end{bmatrix}, \begin{bmatrix} 0 \\ 0 \\ 0 \\ 0 \\ 0 \\ 1 \end{bmatrix} \right\}.$$

Thus $d_1(0) = 3, d_2(0) = 5, d_3(0) = 6$, so $d_3^{ex}(0) = 1, d_2^{ex}(0) = 2, d_1^{ex}(0) = 3$, and $d_3^{new}(0) = 1, d_2^{new}(0) = 1, d_1^{new}(0) = 1$. Also, $d_1(1) = d_1^{ex}(1) = d_1^{new}(1) = 1$.

Thus $m_A(x) = x^3$ and A has ℓESP

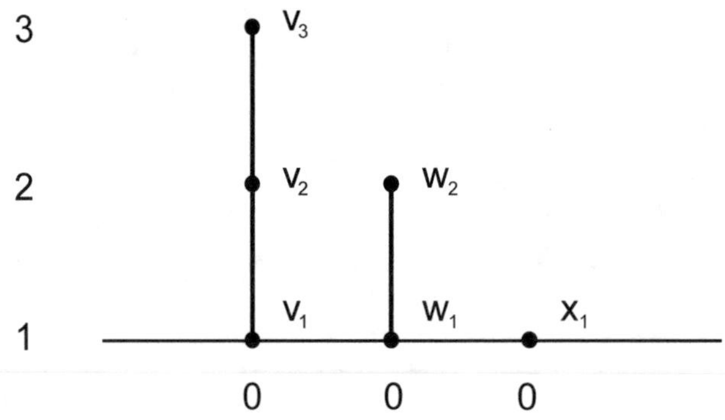

Figure 3.20.

Now we must find the labels. We must choose v_3 to be a vector that is in $E_3(0)$ but that is not

in $E_2(0)$. Given the bases we have found above, we choose $v_3 = \begin{bmatrix} 0 \\ 0 \\ 1 \\ 0 \\ 0 \\ 0 \end{bmatrix}$ and then $v_2 = (A - 0I)v_3 =$

$\begin{bmatrix} 0 \\ 2 \\ 2 \\ -1 \\ -1 \\ 2 \end{bmatrix}$. Now we must choose w_2 in $E_2(3)$ so that $\{v_2, w_2, s_1, t_1, u_1\}$ is linearly independent. We

choose $w_2 = \begin{bmatrix} 1 \\ 0 \\ 0 \\ 0 \\ 0 \\ 0 \end{bmatrix}$. Then $v_1 = (A - 0I)v_2 = \begin{bmatrix} 1 \\ 0 \\ 1 \\ 2 \\ 1 \\ 1 \end{bmatrix}$ and $w_1 = (A - 0I)w_2 = \begin{bmatrix} 2 \\ -1 \\ 1 \\ 5 \\ 3 \\ 1 \end{bmatrix}$. We must

choose x_1 to be a vector in $E_1(3)$ such that $\{v_1, w_1, x_1\}$ is linearly independent. We choose $x_1 =$

$\begin{bmatrix} 1 \\ 0 \\ 0 \\ 0 \\ -1 \\ 0 \end{bmatrix}$.

Then $A = PJP^{-1}$ with

$$P = \begin{bmatrix} 1 & 0 & 0 & 2 & 1 & 1 \\ 0 & 2 & 0 & -1 & 0 & 0 \\ 1 & 2 & 1 & 1 & 0 & 0 \\ 2 & -1 & 0 & 5 & 0 & 0 \\ 1 & -1 & 0 & 3 & 0 & -1 \\ 1 & 2 & 0 & 1 & 0 & 0 \end{bmatrix} \text{ and } J = \begin{bmatrix} 0 & 1 & 0 & 0 & 0 & 0 \\ 0 & 0 & 1 & 0 & 0 & 0 \\ 0 & 0 & 0 & 0 & 0 & 0 \\ 0 & 0 & 0 & 0 & 1 & 0 \\ 0 & 0 & 0 & 0 & 0 & 0 \\ 0 & 0 & 0 & 0 & 0 & 0 \end{bmatrix}.$$

3.5 EXERCISES

For each matrix A, find an invertible matrix P and a matrix in Jordan Canonical Form J with $A = PJP^{-1}$.

1. $A = \begin{bmatrix} 75 & 56 \\ -90 & -67 \end{bmatrix}$, $\quad c_A(x) = (x-3)(x-5)$.

2. $A = \begin{bmatrix} -50 & 99 \\ -20 & 39 \end{bmatrix}$, $\quad c_A(x) = (x+6)(x+5)$.

3. $A = \begin{bmatrix} -18 & 9 \\ -49 & 24 \end{bmatrix}$, $\quad c_A(x) = (x-3)^2$.

4. $A = \begin{bmatrix} 1 & 1 \\ -16 & 9 \end{bmatrix}$, $\quad c_A(x) = (x-5)^2$.

5. $A = \begin{bmatrix} 2 & 1 \\ -25 & 12 \end{bmatrix}$, $\quad c_A(x) = (x-7)^2$.

6. $A = \begin{bmatrix} -15 & 9 \\ -25 & 15 \end{bmatrix}$, $\quad c_A(x) = x^2$.

7. $A = \begin{bmatrix} 1 & 0 & 0 \\ 1 & 2 & -3 \\ 1 & -1 & 0 \end{bmatrix}$, $\quad c_A(x) = (x+1)(x-1)(x-3)$.

8. $A = \begin{bmatrix} 3 & 0 & 2 \\ 1 & 3 & 1 \\ 0 & 1 & 1 \end{bmatrix}$, $\quad c_A(x) = (x-1)(x-2)(x-4)$.

9. $A = \begin{bmatrix} 5 & 8 & 16 \\ 4 & 1 & 8 \\ -4 & -4 & -11 \end{bmatrix}$, $c_A(x) = (x+3)^2(x-1)$.

10. $A = \begin{bmatrix} 4 & 2 & 3 \\ -1 & 1 & -3 \\ 2 & 4 & 9 \end{bmatrix}$, $c_A(x) = (x-3)^2(x-8)$.

11. $A = \begin{bmatrix} 5 & 2 & 1 \\ -1 & 2 & -1 \\ -1 & -2 & 3 \end{bmatrix}$, $c_A(x) = (x-4)^2(x-2)$.

12. $A = \begin{bmatrix} 8 & -3 & -3 \\ 4 & 0 & -2 \\ -2 & 1 & 3 \end{bmatrix}$, $c_A(x) = (x-2)^2(x-7)$.

13. $A = \begin{bmatrix} -3 & 1 & -1 \\ -7 & 5 & -1 \\ -6 & 6 & -2 \end{bmatrix}$, $c_A(x) = (x+2)^2(x-4)$.

14. $A = \begin{bmatrix} 3 & 0 & 0 \\ 9 & -5 & -18 \\ -4 & 4 & 12 \end{bmatrix}$, $c_A(x) = (x-3)^2(x-4)$.

15. $A = \begin{bmatrix} -6 & 9 & 0 \\ -6 & 6 & -2 \\ 9 & -9 & 3 \end{bmatrix}$, $c_A(x) = x^2(x-3)$.

16. $A = \begin{bmatrix} -18 & 42 & 168 \\ 1 & -7 & -40 \\ -2 & 6 & 27 \end{bmatrix}$, $c_A(x) = (x-3)^2(x+4)$.

17. $A = \begin{bmatrix} -1 & 1 & -1 \\ -10 & 6 & -5 \\ -6 & 3 & -2 \end{bmatrix}$, $c_A(x) = (x-1)^3$.

18. $A = \begin{bmatrix} 0 & -4 & 1 \\ 2 & -6 & 1 \\ 4 & -8 & 0 \end{bmatrix}$, $c_A(x) = (x+2)^3$.

19. $A = \begin{bmatrix} -4 & 1 & 2 \\ -5 & 1 & 3 \\ -7 & 2 & 3 \end{bmatrix}$, $c_A(x) = x^3$.

20. $A = \begin{bmatrix} -4 & -2 & 5 \\ -1 & -1 & 1 \\ -2 & -1 & 2 \end{bmatrix}$, $c_A(x) = (x+1)^3$.

21. $A = \begin{bmatrix} -1 & -12 & 2 & 16 \\ -1 & 1 & 1 & 2 \\ -6 & -11 & 7 & 15 \\ -1 & -6 & 1 & 9 \end{bmatrix}$, $c_A(x) = (x-1)(x-3)(x-5)(x-7)$.

22. $A = \begin{bmatrix} 4 & -3 & -4 & 3 \\ 4 & -3 & -8 & 7 \\ -4 & 4 & 6 & -4 \\ -4 & 4 & 2 & 0 \end{bmatrix}$, $c_A(x) = x(x-1)(x-2)(x-4)$.

23. $A = \begin{bmatrix} -8 & -34 & 30 & 6 \\ -9 & -33 & 30 & 6 \\ -7 & -38 & 33 & 6 \\ -36 & -72 & 72 & 18 \end{bmatrix}$, $c_A(x) = x(x-1)(x-3)(x-6)$.

24. $A = \begin{bmatrix} -5 & 2 & -14 & 14 \\ -4 & 1 & -14 & 14 \\ -6 & 2 & -7 & 8 \\ -6 & 2 & -11 & 12 \end{bmatrix}$, $c_A(x) = (x+3)(x+1)(x-1)(x-4)$.

25. $A = \begin{bmatrix} 6 & 1 & -1 & 1 \\ 1 & 6 & -1 & 1 \\ -3 & 1 & 8 & -1 \\ -4 & 0 & 4 & 3 \end{bmatrix}$, $c_A(x) = (x-5)^2(x-6)(x-7)$.

26. $A = \begin{bmatrix} -4 & -3 & 0 & 9 \\ 2 & 2 & 1 & -2 \\ 2 & 0 & 3 & -2 \\ -6 & -3 & 0 & 11 \end{bmatrix}$, $c_A(x) = (x-2)^2(x-3)(x-5)$.

27. $A = \begin{bmatrix} -5 & -2 & 1 & 2 \\ -12 & -7 & 5 & 6 \\ -2 & -1 & 0 & 1 \\ -26 & -13 & 7 & 12 \end{bmatrix}$, $c_A(x) = (x+1)^2 x(x-2)$.

28. $A = \begin{bmatrix} -1 & 3 & -11 & 14 \\ 0 & 5 & -12 & 12 \\ -2 & 3 & -10 & 14 \\ -2 & 3 & -11 & 15 \end{bmatrix}$, $c_A(x) = (x-1)^2(x-2)(x-5)$.

29. $A = \begin{bmatrix} 8 & -1 & -8 & 12 \\ -6 & 4 & 12 & -17 \\ -6 & 2 & 16 & -19 \\ -6 & 2 & 12 & -15 \end{bmatrix}$, $c_A(x) = (x-2)^2(x-4)(x-5)$.

30. $A = \begin{bmatrix} 4 & 2 & 1 & -6 \\ -16 & 4 & -16 & 0 \\ -6 & -1 & -3 & 4 \\ -6 & 1 & -6 & 2 \end{bmatrix}$, $c_A(x) = x^2(x-3)(x-4)$.

31. $A = \begin{bmatrix} 10 & 5 & -2 & -17 \\ 8 & 8 & -4 & -18 \\ 0 & 6 & -4 & -6 \\ 8 & 3 & -1 & -13 \end{bmatrix}$, $c_A(x) = (x-2)^2(x+1)(x+2)$.

32. $A = \begin{bmatrix} 6 & -6 & -8 & 3 \\ -4 & 1 & 8 & 1 \\ 4 & -4 & -6 & 2 \\ -4 & -2 & 8 & 4 \end{bmatrix}$, $c_A(x) = x^2(x-2)(x-3)$.

33. $A = \begin{bmatrix} 22 & -5 & 1 & -5 \\ 18 & -2 & 2 & -5 \\ 16 & -4 & 3 & -4 \\ 62 & -16 & 2 & -13 \end{bmatrix}$, $c_A(x) = (x-2)^2(x-3)^2$.

34. $A = \begin{bmatrix} 5 & 0 & 2 & 1 \\ 0 & 2 & 1 & 1 \\ -2 & 0 & 0 & -2 \\ 2 & -2 & 5 & 7 \end{bmatrix}$, $c_A(x) = (x-3)^2(x-4)^2$.

35. $A = \begin{bmatrix} -5 & 2 & -10 & 10 \\ -1 & 2 & -1 & 1 \\ -5 & 1 & 0 & 5 \\ -10 & 2 & -10 & 15 \end{bmatrix}$, $c_A(x) = (x - 1)^2(x - 5)^2$.

36. $A = \begin{bmatrix} -1 & -2 & -1 & 3 \\ -6 & -5 & 1 & 6 \\ -6 & -4 & 0 & 6 \\ -6 & -7 & 1 & 8 \end{bmatrix}$, $c_A(x) = (x + 1)^2(x - 2)^2$.

37. $A = \begin{bmatrix} -3 & -4 & 0 & 6 \\ -1 & -2 & 1 & 2 \\ -1 & -2 & 1 & 2 \\ -3 & -5 & 1 & 6 \end{bmatrix}$, $c_A(x) = x^2(x - 1)^2$.

38. $A = \begin{bmatrix} -1 & 1 & -2 & 0 \\ -10 & 1 & 2 & 2 \\ -5 & 1 & 0 & 1 \\ -6 & 2 & -4 & 2 \end{bmatrix}$, $c_A(x) = (x + 1)^2(x - 2)^2$.

39. $A = \begin{bmatrix} -5 & 3 & -3 & 6 \\ -4 & 3 & -2 & 4 \\ -4 & 2 & -1 & 4 \\ -8 & 3 & -3 & 9 \end{bmatrix}$, $c_A(x) = (x - 1)^3(x - 3)$.

40. $A = \begin{bmatrix} 8 & 4 & -3 & 1 \\ 12 & 8 & -6 & 6 \\ 24 & 14 & -10 & 8 \\ 6 & 3 & -3 & 5 \end{bmatrix}$, $c_A(x) = (x - 2)^3(x - 5)$.

41. $A = \begin{bmatrix} 7 & 10 & -6 & -28 \\ 2 & 4 & -2 & -9 \\ 2 & 3 & -1 & -9 \\ 2 & 3 & -2 & -8 \end{bmatrix}$, $c_A(x) = (x - 1)^3(x + 1)$.

42. $A = \begin{bmatrix} 4 & 6 & -5 & -1 \\ 5 & 4 & -5 & 0 \\ 5 & 6 & -6 & -1 \\ 5 & 5 & -5 & -1 \end{bmatrix}$, $c_A(x) = (x + 1)^3(x - 4)$.

43. $A = \begin{bmatrix} 0 & 1 & -1 & 1 \\ 0 & 0 & 1 & 0 \\ 0 & -1 & 2 & 0 \\ -2 & 1 & -2 & 3 \end{bmatrix}$, $c_A(x) = (x-1)^3(x-2)$.

44. $A = \begin{bmatrix} -12 & 15 & 7 & -11 \\ -14 & 16 & 7 & -10 \\ -12 & 14 & 8 & -11 \\ -12 & 12 & 6 & -7 \end{bmatrix}$, $c_A(x) = (x-2)^3(x+1)$.

45. $A = \begin{bmatrix} 5 & 2 & -1 & -1 \\ 3 & 10 & -3 & -3 \\ 4 & 8 & 0 & -4 \\ 3 & 6 & -3 & 1 \end{bmatrix}$, $c_A(x) = (x-4)^4$.

46. $A = \begin{bmatrix} -5 & 4 & -2 & 2 \\ -8 & 7 & -4 & 4 \\ -10 & 10 & -6 & 5 \\ -2 & 2 & -1 & 0 \end{bmatrix}$, $c_A(x) = (x+1)^4$.

47. $A = \begin{bmatrix} 4 & 1 & 2 & -2 \\ -1 & 2 & 2 & 0 \\ 0 & 0 & 5 & -1 \\ 1 & 1 & 2 & 1 \end{bmatrix}$, $c_A(x) = (x-3)^4$.

48. $A = \begin{bmatrix} 4 & 4 & -2 & -2 \\ -1 & 7 & 0 & -1 \\ -1 & 1 & 6 & -1 \\ 1 & -2 & 1 & 7 \end{bmatrix}$, $c_A(x) = (x-6)^4$.

49. $A = \begin{bmatrix} 5 & -1 & 2 & 4 \\ 5 & -1 & 3 & 4 \\ 8 & -5 & 3 & 4 \\ -4 & 1 & -1 & -3 \end{bmatrix}$, $c_A(x) = (x-1)^4$.

50. $A = \begin{bmatrix} -2 & 2 & 3 & 3 \\ -4 & 4 & 4 & 3 \\ -1 & -1 & 4 & 3 \\ -2 & 2 & 1 & 2 \end{bmatrix}$, $c_A(x) = (x-2)^4$.

51. $A = \begin{bmatrix} 2 & -3 & -2 & 3 \\ -4 & 2 & 4 & -4 \\ -4 & 0 & 4 & -3 \\ -8 & 4 & 8 & -8 \end{bmatrix}$, $c_A(x) = x^4$.

52. $A = \begin{bmatrix} 1 & 3 & -6 & 0 \\ -2 & 3 & -4 & 6 \\ -1 & 1 & -1 & 3 \\ 0 & 1 & -2 & 1 \end{bmatrix}$, $c_A(x) = (x-1)^4$.

53. $A = \begin{bmatrix} -5 & -2 & 0 & -4 & 8 \\ 2 & 1 & -2 & 2 & 0 \\ -3 & -1 & -2 & -3 & 8 \\ -1 & -1 & 1 & -2 & 0 \\ -4 & -1 & 0 & -2 & 7 \end{bmatrix}$, $c_A(x) = (x+1)^4(x-3)$.

54. $A = \begin{bmatrix} -1 & 1 & -4 & 2 & 4 \\ -1 & -1 & 2 & 2 & 0 \\ -1 & -1 & 2 & 2 & 0 \\ 0 & 1 & -3 & 0 & 2 \\ -1 & -1 & 0 & 2 & 2 \end{bmatrix}$, $c_A(x) = x^4(x-2)$.

55. $A = \begin{bmatrix} 5 & 1 & 3 & -1 & -3 \\ 3 & 1 & 4 & 1 & -3 \\ 0 & -4 & 4 & 4 & 0 \\ 3 & 1 & 3 & 1 & -3 \\ 6 & -3 & 8 & 3 & -4 \end{bmatrix}$, $c_A(x) = (x-2)^4(x+1)$.

56. $A = \begin{bmatrix} 5 & 1 & 0 & 0 & -2 \\ 1 & 6 & 1 & -2 & -2 \\ 1 & 2 & 5 & -2 & -2 \\ 1 & 2 & 1 & 2 & -2 \\ 2 & 1 & 0 & 0 & 1 \end{bmatrix}$, $c_A(x) = (x-4)^4(x-3)$.

57. $A = \begin{bmatrix} -2 & -9 & -10 & -7 & 10 \\ 8 & 21 & 21 & 15 & -20 \\ -2 & -1 & 1 & 0 & 0 \\ -8 & -18 & -20 & -12 & 20 \\ -2 & -5 & -5 & -4 & 7 \end{bmatrix}$, $c_A(x) = (x-2)^4(x-7)$.

58. $A = \begin{bmatrix} -1 & 2 & 4 & 2 & -1 \\ -3 & 5 & 4 & 1 & -1 \\ -1 & 0 & 4 & 1 & 0 \\ -3 & 2 & 3 & 4 & -1 \\ -2 & 1 & 3 & 1 & 2 \end{bmatrix}$, $c_A(x) = (x-3)^4(x-2)$.

59. $A = \begin{bmatrix} 5 & 7 & 5 & 2 & -2 & -8 \\ 2 & 10 & 7 & 2 & -3 & -7 \\ -2 & -7 & -4 & -2 & 3 & 7 \\ 0 & 0 & -1 & 3 & 1 & 0 \\ -2 & -7 & -7 & -2 & 6 & 7 \\ 2 & 6 & 5 & 2 & -2 & -4 \end{bmatrix}$, $c_A(x) = (x-3)^5(x-1)$.

60. $A = \begin{bmatrix} 9 & 3 & 0 & 3 & -3 & -12 \\ 0 & 3 & 3 & 1 & -2 & 0 \\ 0 & -2 & 8 & 1 & -2 & 0 \\ 0 & -1 & 0 & 4 & 2 & 0 \\ 0 & -2 & 2 & 0 & 5 & 0 \\ 2 & 1 & 0 & 1 & -1 & -1 \end{bmatrix}$, $c_A(x) = (x-5)^5(x-3)$.

APPENDIX A

Answers to odd-numbered exercises

A.1 ANSWERS TO EXERCISES–CHAPTER 2

1. $J = \begin{bmatrix} 2 & 0 & 0 & 0 \\ 0 & 3 & 0 & 0 \\ 0 & 0 & 5 & 0 \\ 0 & 0 & 0 & 7 \end{bmatrix}$, $\quad m_A(x) = (x-2)(x-3)(x-5)(x-7)$,

alg-mult(2) = 1, geom-mult(2) = 1, max-ind(2) = 1,
alg-mult(3) = 1, geom-mult(3) = 1, max-ind(3) = 1,
alg-mult(5) = 1, geom-mult(5) = 1, max-ind(5) = 1,
alg-mult(7) = 1, geom-mult(7) = 1, max-ind(7) = 1.

3. $J = \begin{bmatrix} 2 & 1 & 0 & 0 & 0 \\ 0 & 2 & 1 & 0 & 0 \\ 0 & 0 & 2 & 0 & 0 \\ 0 & 0 & 0 & 3 & 1 \\ 0 & 0 & 0 & 0 & 3 \end{bmatrix}$, $\quad m_A(x) = (x-2)^3(x-3)^2$,

alg-mult(2) = 3, geom-mult(2) = 1, max-ind(2) = 3,
alg-mult(3) = 2, geom-mult(3) = 1, max-ind(3) = 2;

$J = \begin{bmatrix} 2 & 1 & 0 & 0 & 0 \\ 0 & 2 & 1 & 0 & 0 \\ 0 & 0 & 2 & 0 & 0 \\ 0 & 0 & 0 & 3 & 0 \\ 0 & 0 & 0 & 0 & 3 \end{bmatrix}$, $\quad m_A(x) = (x-2)^3(x-3)$,

alg-mult(2) = 3, geom-mult(2) = 1, max-ind(2) = 3,
alg-mult(3) = 2, geom-mult(3) = 2, max-ind(3) = 1;

$J = \begin{bmatrix} 2 & 1 & 0 & 0 & 0 \\ 0 & 2 & 0 & 0 & 0 \\ 0 & 0 & 2 & 0 & 0 \\ 0 & 0 & 0 & 3 & 1 \\ 0 & 0 & 0 & 0 & 3 \end{bmatrix}$, $\quad m_A(x) = (x-2)^2(x-3)^2$,

alg-mult(2) = 3, geom-mult(2) = 2, max-ind(2) = 2,
alg-mult(3) = 2, geom-mult(3) = 1, max-ind(3) = 2;

$$J = \begin{bmatrix} 2 & 1 & 0 & 0 & 0 \\ 0 & 2 & 0 & 0 & 0 \\ 0 & 0 & 2 & 0 & 0 \\ 0 & 0 & 0 & 3 & 0 \\ 0 & 0 & 0 & 0 & 3 \end{bmatrix}, \qquad m_A(x) = (x-2)^2(x-3)^2,$$

alg-mult(2) = 3, geom-mult(2) = 2, max-ind(2) = 2,
alg-mult(3) = 2, geom-mult(3) = 2, max-ind(3) = 1;

$$J = \begin{bmatrix} 2 & 0 & 0 & 0 & 0 \\ 0 & 2 & 0 & 0 & 0 \\ 0 & 0 & 2 & 0 & 0 \\ 0 & 0 & 0 & 3 & 1 \\ 0 & 0 & 0 & 0 & 3 \end{bmatrix}, \qquad m_A(x) = (x-2)(x-3)^2,$$

alg-mult(2) = 3, geom-mult(2) = 3, max-ind(2) = 1,
alg-mult(3) = 2, geom-mult(3) = 1, max-ind(3) = 2;

$$J = \begin{bmatrix} 2 & 0 & 0 & 0 & 0 \\ 0 & 2 & 0 & 0 & 0 \\ 0 & 0 & 2 & 0 & 0 \\ 0 & 0 & 0 & 3 & 0 \\ 0 & 0 & 0 & 0 & 3 \end{bmatrix}, \qquad m_A(x) = (x-2)(x-3),$$

alg-mult(2) = 3, geom-mult(2) = 3, max-ind(2) = 1,
alg-mult(3) = 2, geom-mult(3) = 2, max-ind(3) = 1.

5. $J = \begin{bmatrix} 3 & 1 & 0 & 0 & 0 & 0 & 0 & 0 \\ 0 & 3 & 0 & 0 & 0 & 0 & 0 & 0 \\ 0 & 0 & 3 & 1 & 0 & 0 & 0 & 0 \\ 0 & 0 & 0 & 3 & 0 & 0 & 0 & 0 \\ 0 & 0 & 0 & 0 & 5 & 1 & 0 & 0 \\ 0 & 0 & 0 & 0 & 0 & 5 & 0 & 0 \\ 0 & 0 & 0 & 0 & 0 & 0 & 5 & 1 \\ 0 & 0 & 0 & 0 & 0 & 0 & 0 & 5 \end{bmatrix},$

alg-mult(3) = 4, geom-mult(3) = 2, max-ind(3) = 2,
alg-mult(5) = 4, geom-mult(5) = 2, max-ind(5) = 2;

$$J = \begin{bmatrix} 3 & 1 & 0 & 0 & 0 & 0 & 0 & 0 \\ 0 & 3 & 0 & 0 & 0 & 0 & 0 & 0 \\ 0 & 0 & 3 & 1 & 0 & 0 & 0 & 0 \\ 0 & 0 & 0 & 3 & 0 & 0 & 0 & 0 \\ 0 & 0 & 0 & 0 & 5 & 1 & 0 & 0 \\ 0 & 0 & 0 & 0 & 0 & 5 & 0 & 0 \\ 0 & 0 & 0 & 0 & 0 & 0 & 5 & 0 \\ 0 & 0 & 0 & 0 & 0 & 0 & 0 & 5 \end{bmatrix},$$

alg-mult(3) = 4, geom-mult(3) = 2, max-ind(3) = 2,
alg-mult(5) = 4, geom-mult(5) = 3, max-ind(5) = 2;

$$J = \begin{bmatrix} 3 & 1 & 0 & 0 & 0 & 0 & 0 & 0 \\ 0 & 3 & 0 & 0 & 0 & 0 & 0 & 0 \\ 0 & 0 & 3 & 0 & 0 & 0 & 0 & 0 \\ 0 & 0 & 0 & 3 & 0 & 0 & 0 & 0 \\ 0 & 0 & 0 & 0 & 5 & 1 & 0 & 0 \\ 0 & 0 & 0 & 0 & 0 & 5 & 0 & 0 \\ 0 & 0 & 0 & 0 & 0 & 0 & 5 & 1 \\ 0 & 0 & 0 & 0 & 0 & 0 & 0 & 5 \end{bmatrix},$$

alg-mult(3) = 4, geom-mult(3) = 3, max-ind(3) = 2,
alg-mult(5) = 4, geom-mult(5) = 2, max-ind(5) = 2;

$$J = \begin{bmatrix} 3 & 1 & 0 & 0 & 0 & 0 & 0 & 0 \\ 0 & 3 & 0 & 0 & 0 & 0 & 0 & 0 \\ 0 & 0 & 3 & 0 & 0 & 0 & 0 & 0 \\ 0 & 0 & 0 & 3 & 0 & 0 & 0 & 0 \\ 0 & 0 & 0 & 0 & 5 & 1 & 0 & 0 \\ 0 & 0 & 0 & 0 & 0 & 5 & 0 & 0 \\ 0 & 0 & 0 & 0 & 0 & 0 & 5 & 0 \\ 0 & 0 & 0 & 0 & 0 & 0 & 0 & 5 \end{bmatrix},$$

alg-mult(3) = 4, geom-mult(3) = 3, max-ind(3) = 2,
alg-mult(5) = 4, geom-mult(5) = 3, max-ind(5) = 2.

7. $J = \begin{bmatrix} 0 & 0 & 0 & 0 & 0 & 0 & 0 & 0 \\ 0 & 0 & 0 & 0 & 0 & 0 & 0 & 0 \\ 0 & 0 & 2 & 1 & 0 & 0 & 0 & 0 \\ 0 & 0 & 0 & 2 & 0 & 0 & 0 & 0 \\ 0 & 0 & 0 & 0 & 2 & 1 & 0 & 0 \\ 0 & 0 & 0 & 0 & 0 & 2 & 0 & 0 \\ 0 & 0 & 0 & 0 & 0 & 0 & 6 & 1 \\ 0 & 0 & 0 & 0 & 0 & 0 & 0 & 6 \end{bmatrix}$, $m_A(x) = x(x-2)^2(x-6)^2$,

alg-mult$(0) = 2$, geom-mult$(0) = 1$,
alg-mult$(2) = 4$, geom-mult$(2) = 2$,
alg-mult$(6) = 2$, geom-mult$(6) = 1$;

$J = \begin{bmatrix} 0 & 0 & 0 & 0 & 0 & 0 & 0 & 0 \\ 0 & 0 & 0 & 0 & 0 & 0 & 0 & 0 \\ 0 & 0 & 2 & 1 & 0 & 0 & 0 & 0 \\ 0 & 0 & 0 & 2 & 0 & 0 & 0 & 0 \\ 0 & 0 & 0 & 0 & 2 & 0 & 0 & 0 \\ 0 & 0 & 0 & 0 & 0 & 2 & 0 & 0 \\ 0 & 0 & 0 & 0 & 0 & 0 & 6 & 1 \\ 0 & 0 & 0 & 0 & 0 & 0 & 0 & 6 \end{bmatrix}$, $m_A(x) = x(x-2)^2(x-6)^2$,

alg-mult$(0) = 2$, geom-mult$(0) = 1$,
alg-mult$(2) = 4$, geom-mult$(2) = 3$,
alg-mult$(6) = 2$, geom-mult$(6) = 1$.

9. $J = \begin{bmatrix} 3 & 1 & 0 & 0 & 0 & 0 \\ 0 & 3 & 1 & 0 & 0 & 0 \\ 0 & 0 & 3 & 0 & 0 & 0 \\ 0 & 0 & 0 & 3 & 0 & 0 \\ 0 & 0 & 0 & 0 & 8 & 1 \\ 0 & 0 & 0 & 0 & 0 & 8 \end{bmatrix}$, $m_A(x) = (x-3)^3(x-8)^2$,

alg-mult$(3) = 4$, max-ind$(3) = 3$,
alg-mult$(8) = 2$, max-ind$(8) = 2$;

$J = \begin{bmatrix} 3 & 1 & 0 & 0 & 0 & 0 \\ 0 & 3 & 0 & 0 & 0 & 0 \\ 0 & 0 & 3 & 1 & 0 & 0 \\ 0 & 0 & 0 & 3 & 0 & 0 \\ 0 & 0 & 0 & 0 & 8 & 1 \\ 0 & 0 & 0 & 0 & 0 & 8 \end{bmatrix}$, $m_A(x) = (x-3)^2(x-8)^2$,

alg-mult$(3) = 4$, max-ind$(3) = 2$,

alg-mult(8) = 2, max-ind(8) = 2.

11. $J = \begin{bmatrix} 1 & 1 & 0 & 0 & 0 & 0 \\ 0 & 1 & 1 & 0 & 0 & 0 \\ 0 & 0 & 1 & 1 & 0 & 0 \\ 0 & 0 & 0 & 1 & 0 & 0 \\ 0 & 0 & 0 & 0 & 1 & 0 \\ 0 & 0 & 0 & 0 & 0 & 4 \end{bmatrix},$ $c_A(x) = (x - 1)^5(x - 4),$

alg-mult(1) = 5, geom-mult(1) = 2, max-ind(1) = 4,
alg-mult(4) = 1, geom-mult(4) = 1, max-ind(4) = 1;

$J = \begin{bmatrix} 1 & 1 & 0 & 0 & 0 & 0 \\ 0 & 1 & 1 & 0 & 0 & 0 \\ 0 & 0 & 1 & 1 & 0 & 0 \\ 0 & 0 & 0 & 1 & 0 & 0 \\ 0 & 0 & 0 & 0 & 4 & 0 \\ 0 & 0 & 0 & 0 & 0 & 4 \end{bmatrix},$ $c_A(x) = (x - 1)^4(x - 4)^2,$

alg-mult(1) = 4, geom-mult(1) = 1, max-ind(1) = 4,
alg-mult(4) = 2, geom-mult(4) = 2, max-ind(4) = 1.

13. $J = \begin{bmatrix} 1 & 1 & 0 & 0 & 0 \\ 0 & 1 & 1 & 0 & 0 \\ 0 & 0 & 1 & 0 & 0 \\ 0 & 0 & 0 & 1 & 0 \\ 0 & 0 & 0 & 0 & 1 \end{bmatrix},$ $c_A(x) = (x - 1)^5, m_A(x) = (x - 1)^3,$

max-ind(1) = 3;

$J = \begin{bmatrix} 1 & 1 & 0 & 0 & 0 \\ 0 & 1 & 0 & 0 & 0 \\ 0 & 0 & 1 & 1 & 0 \\ 0 & 0 & 0 & 1 & 0 \\ 0 & 0 & 0 & 0 & 1 \end{bmatrix},$ $c_A(x) = (x - 1)^5, m_A(x) = (x - 1)^2,$

max-ind(1) = 2.

15. $J = \begin{bmatrix} 3 & 1 & 0 & 0 & 0 \\ 0 & 3 & 0 & 0 & 0 \\ 0 & 0 & 3 & 1 & 0 \\ 0 & 0 & 0 & 3 & 0 \\ 0 & 0 & 0 & 0 & 3 \end{bmatrix}$, $c_A(x) = (x-3)^5, m_A(x) = (x-3)^2,$

geom-mult$(3) = 3$;

$J = \begin{bmatrix} 3 & 1 & 0 & 0 & 0 \\ 0 & 3 & 0 & 0 & 0 \\ 0 & 0 & 3 & 0 & 0 \\ 0 & 0 & 0 & 3 & 0 \\ 0 & 0 & 0 & 0 & 3 \end{bmatrix}$, $c_A(x) = (x-3)^5, m_A(x) = (x-3)^2,$

geom-mult$(3) = 4$.

17. $J = \begin{bmatrix} 8 & 1 & 0 & 0 & 0 & 0 \\ 0 & 8 & 1 & 0 & 0 & 0 \\ 0 & 0 & 8 & 1 & 0 & 0 \\ 0 & 0 & 0 & 8 & 0 & 0 \\ 0 & 0 & 0 & 0 & 8 & 0 \\ 0 & 0 & 0 & 0 & 0 & 8 \end{bmatrix}$, $c_A(x) = (x-8)^6, m_A(x) = (x-8)^4.$

A.2   ANSWERS TO EXERCISES–CHAPTER 3

Note that while J is unique up to the order of the blocks, P is not at all unique, so it is perfectly possible (and indeed quite likely) that you may have a correct value for P that is (much) different than the one given here. You can always check your answer by computing PJP^{-1} and seeing if it is equal to A. That is rather tedious, as it requires you to find P^{-1}. An equivalent but much easier method is: (a) Check that P is invertible; and (b) Check that $AP = PJ$.

1. $P = \begin{bmatrix} -7 & 4 \\ 9 & -5 \end{bmatrix}$, $J = \begin{bmatrix} 3 & 0 \\ 0 & 5 \end{bmatrix}$.

3. $P = \begin{bmatrix} -21 & 1 \\ -49 & 0 \end{bmatrix}$, $J = \begin{bmatrix} 3 & 1 \\ 0 & 3 \end{bmatrix}$.

5. $P = \begin{bmatrix} -5 & 1 \\ -25 & 0 \end{bmatrix}$, $J = \begin{bmatrix} 7 & 1 \\ 0 & 7 \end{bmatrix}$.

7. $P = \begin{bmatrix} 0 & 2 & 0 \\ 1 & 1 & -3 \\ 1 & 1 & 1 \end{bmatrix}$, $J = \begin{bmatrix} -1 & 0 & 0 \\ 0 & 1 & 0 \\ 0 & 0 & 3 \end{bmatrix}$.

9. $P = \begin{bmatrix} -1 & -2 & -2 \\ 1 & 0 & -1 \\ 0 & 1 & 1 \end{bmatrix}$, $J = \begin{bmatrix} -3 & 0 & 0 \\ 0 & -3 & 0 \\ 0 & 0 & 1 \end{bmatrix}$.

11. $P = \begin{bmatrix} -1 & -2 & -1 \\ 0 & 1 & 1 \\ 1 & 0 & 1 \end{bmatrix}$, $J = \begin{bmatrix} 4 & 0 & 0 \\ 0 & 4 & 0 \\ 0 & 0 & 2 \end{bmatrix}$.

13. $P = \begin{bmatrix} -1 & 0 & 0 \\ -1 & 0 & 1 \\ 0 & 1 & 1 \end{bmatrix}$, $J = \begin{bmatrix} -2 & 1 & 0 \\ 0 & -2 & 0 \\ 0 & 0 & 4 \end{bmatrix}$.

15. $P = \begin{bmatrix} -3 & -1 & -2 \\ -2 & -1 & -2 \\ 3 & 1 & 3 \end{bmatrix}$, $J = \begin{bmatrix} 0 & 1 & 0 \\ 0 & 0 & 0 \\ 0 & 0 & 3 \end{bmatrix}$.

17. $P = \begin{bmatrix} -2 & 1 & 1 \\ -10 & 0 & 2 \\ -6 & 0 & 0 \end{bmatrix}$, $J = \begin{bmatrix} 1 & 1 & 0 \\ 0 & 1 & 0 \\ 0 & 0 & 1 \end{bmatrix}$.

19. $P = \begin{bmatrix} 1 & 2 & 0 \\ 2 & 3 & 0 \\ 1 & 3 & 1 \end{bmatrix}$, $J = \begin{bmatrix} 0 & 1 & 0 \\ 0 & 0 & 1 \\ 0 & 0 & 0 \end{bmatrix}$.

21. $P = \begin{bmatrix} 1 & 2 & 2 & 22 \\ 0 & 1 & 1 & 3 \\ 1 & 2 & 4 & 18 \\ 0 & 1 & 1 & 11 \end{bmatrix}$, $J = \begin{bmatrix} 1 & 0 & 0 & 0 \\ 0 & 3 & 0 & 0 \\ 0 & 0 & 5 & 0 \\ 0 & 0 & 0 & 7 \end{bmatrix}$.

23. $P = \begin{bmatrix} 1 & 2 & 2 & 1 \\ 1 & 3 & 2 & 1 \\ 1 & 4 & 3 & 1 \\ 2 & 0 & 0 & 3 \end{bmatrix}$, $J = \begin{bmatrix} 0 & 0 & 0 & 0 \\ 0 & 1 & 0 & 0 \\ 0 & 0 & 3 & 0 \\ 0 & 0 & 0 & 6 \end{bmatrix}$.

25. $P = \begin{bmatrix} 1 & 0 & 1 & 0 \\ 0 & 1 & 1 & 0 \\ 1 & 1 & 1 & 1 \\ 0 & -2 & 0 & 1 \end{bmatrix}$, $J = \begin{bmatrix} 5 & 0 & 0 & 0 \\ 0 & 5 & 0 & 0 \\ 0 & 0 & 6 & 0 \\ 0 & 0 & 0 & 7 \end{bmatrix}$.

27. $P = \begin{bmatrix} 1 & 0 & 1 & 5 \\ 0 & 1 & 5 & 17 \\ 0 & 0 & 1 & 3 \\ 2 & 1 & 7 & 33 \end{bmatrix}$, $J = \begin{bmatrix} -1 & 0 & 0 & 0 \\ 0 & -1 & 0 & 0 \\ 0 & 0 & 0 & 0 \\ 0 & 0 & 0 & 2 \end{bmatrix}$.

29. $P = \begin{bmatrix} -2 & 1 & 2 & 1 \\ 4 & 0 & 0 & -1 \\ 4 & -2 & 1 & -1 \\ 4 & -2 & 0 & -1 \end{bmatrix}$, $J = \begin{bmatrix} 2 & 0 & 0 & 0 \\ 0 & 2 & 0 & 0 \\ 0 & 0 & 4 & 0 \\ 0 & 0 & 0 & 5 \end{bmatrix}$.

31. $P = \begin{bmatrix} 3 & 0 & 1 & 1 \\ 2 & 1 & 2 & 1 \\ 0 & 1 & 2 & 0 \\ 2 & 0 & 1 & 1 \end{bmatrix}$, $J = \begin{bmatrix} 2 & 1 & 0 & 0 \\ 0 & 2 & 0 & 0 \\ 0 & 0 & -1 & 0 \\ 0 & 0 & 0 & -2 \end{bmatrix}$.

33. $P = \begin{bmatrix} 1 & 0 & 1 & 0 \\ 2 & 5 & 4 & 1 \\ 0 & -4 & 1 & 0 \\ 2 & -6 & 0 & -1 \end{bmatrix}$, $J = \begin{bmatrix} 2 & 1 & 0 & 0 \\ 0 & 2 & 0 & 0 \\ 0 & 0 & 3 & 0 \\ 0 & 0 & 0 & 3 \end{bmatrix}$.

35. $P = \begin{bmatrix} 2 & 0 & 1 & 1 \\ 1 & 1 & 0 & 0 \\ 1 & 0 & 1 & 0 \\ 2 & 0 & 2 & 1 \end{bmatrix}$, $J = \begin{bmatrix} 1 & 1 & 0 & 0 \\ 0 & 1 & 0 & 0 \\ 0 & 0 & 5 & 0 \\ 0 & 0 & 0 & 5 \end{bmatrix}$.

37. $P = \begin{bmatrix} 2 & 0 & 2 & 1 \\ 0 & 1 & 1 & 0 \\ 0 & 0 & 1 & 0 \\ 1 & 1 & 2 & 1 \end{bmatrix}$, $J = \begin{bmatrix} 0 & 1 & 0 & 0 \\ 0 & 0 & 0 & 0 \\ 0 & 0 & 1 & 1 \\ 0 & 0 & 0 & 1 \end{bmatrix}$.

39. $P = \begin{bmatrix} 3 & 2 & 1 & 3 \\ 2 & 3 & 1 & 2 \\ 2 & 2 & 1 & 2 \\ 3 & 2 & 1 & 4 \end{bmatrix}$, $J = \begin{bmatrix} 1 & 1 & 0 & 0 \\ 0 & 1 & 0 & 0 \\ 0 & 0 & 1 & 0 \\ 0 & 0 & 0 & 3 \end{bmatrix}$.

41. $P = \begin{bmatrix} 4 & 3 & 5 & 3 \\ 1 & 2 & 1 & 1 \\ 1 & 1 & 2 & 1 \\ 1 & 1 & 1 & 1 \end{bmatrix}$, $J = \begin{bmatrix} 1 & 1 & 0 & 0 \\ 0 & 1 & 0 & 0 \\ 0 & 0 & 1 & 0 \\ 0 & 0 & 0 & -1 \end{bmatrix}$.

43. $P = \begin{bmatrix} 1 & 1 & 0 & 1 \\ 0 & 1 & 0 & 0 \\ 0 & 1 & 1 & 0 \\ 1 & 2 & 2 & 2 \end{bmatrix}$, $J = \begin{bmatrix} 1 & 1 & 0 & 0 \\ 0 & 1 & 1 & 0 \\ 0 & 0 & 1 & 0 \\ 0 & 0 & 0 & 2 \end{bmatrix}$.

45. $P = \begin{bmatrix} 1 & 2 & 1 & 0 \\ 3 & 4 & 2 & 1 \\ 4 & 7 & 4 & 1 \\ 3 & 2 & 1 & 1 \end{bmatrix}$, $J = \begin{bmatrix} 4 & 1 & 0 & 0 \\ 0 & 4 & 0 & 0 \\ 0 & 0 & 4 & 0 \\ 0 & 0 & 0 & 4 \end{bmatrix}$.

47. $P = \begin{bmatrix} 2 & 1 & 1 & 3 \\ 0 & 1 & 0 & 1 \\ 1 & 1 & 1 & 2 \\ 2 & 1 & 1 & 4 \end{bmatrix}$, $J = \begin{bmatrix} 3 & 1 & 0 & 0 \\ 0 & 3 & 1 & 0 \\ 0 & 0 & 3 & 0 \\ 0 & 0 & 0 & 3 \end{bmatrix}$.

49. $P = \begin{bmatrix} 4 & 1 & 1 & 3 \\ 4 & 2 & 1 & 3 \\ 0 & 1 & 1 & 0 \\ -3 & 0 & -1 & -2 \end{bmatrix}$, $J = \begin{bmatrix} 1 & 1 & 0 & 0 \\ 0 & 1 & 1 & 0 \\ 0 & 0 & 1 & 1 \\ 0 & 0 & 0 & 1 \end{bmatrix}$.

51. $P = \begin{bmatrix} -1 & -2 & 1 & -1 \\ 2 & 1 & 0 & 0 \\ 2 & 0 & 1 & 0 \\ 4 & 1 & 0 & 1 \end{bmatrix}$, $J = \begin{bmatrix} 0 & 1 & 0 & 0 \\ 0 & 0 & 0 & 0 \\ 0 & 0 & 0 & 1 \\ 0 & 0 & 0 & 0 \end{bmatrix}$.

53. $P = \begin{bmatrix} 2 & 0 & 0 & 0 & 1 \\ 0 & 1 & 2 & 2 & 0 \\ 2 & 0 & 1 & 0 & 1 \\ 0 & -1 & -1 & -1 & 0 \\ 1 & 0 & 0 & 0 & 1 \end{bmatrix}$, $J = \begin{bmatrix} -1 & 1 & 0 & 0 & 0 \\ 0 & -1 & 0 & 0 & 0 \\ 0 & 0 & -1 & 1 & 0 \\ 0 & 0 & 0 & -1 & 0 \\ 0 & 0 & 0 & 0 & 3 \end{bmatrix}$.

55. $P = \begin{bmatrix} 1 & 0 & 0 & 0 & 1 \\ 1 & 1 & 0 & 1 & 1 \\ 0 & 2 & 1 & 0 & 0 \\ 1 & 0 & 0 & 1 & 1 \\ 1 & 2 & 1 & 0 & 2 \end{bmatrix}$, $\quad J = \begin{bmatrix} 2 & 1 & 0 & 0 & 0 \\ 0 & 2 & 1 & 0 & 0 \\ 0 & 0 & 2 & 0 & 0 \\ 0 & 0 & 0 & 2 & 0 \\ 0 & 0 & 0 & 0 & -1 \end{bmatrix}$.

57. $P = \begin{bmatrix} 1 & 0 & 2 & 0 & 2 \\ -2 & 1 & -4 & -1 & -4 \\ 0 & -1 & 1 & 0 & 0 \\ 2 & 0 & 4 & 1 & 4 \\ 0 & 0 & 1 & 0 & 1 \end{bmatrix}$, $\quad J = \begin{bmatrix} 2 & 1 & 0 & 0 & 0 \\ 0 & 2 & 1 & 0 & 0 \\ 0 & 0 & 2 & 1 & 0 \\ 0 & 0 & 0 & 2 & 0 \\ 0 & 0 & 0 & 0 & 7 \end{bmatrix}$.

59. $P = \begin{bmatrix} 2 & 1 & 1 & 1 & 1 & 1 \\ 1 & 2 & 1 & 1 & 1 & 1 \\ -1 & -2 & 0 & -1 & 0 & -1 \\ 0 & 0 & 0 & 1 & 1 & 0 \\ -1 & -2 & 0 & -1 & 1 & -1 \\ 1 & 1 & 1 & 1 & 1 & 1 \end{bmatrix}$, $\quad J = \begin{bmatrix} 3 & 1 & 0 & 0 & 0 & 0 \\ 0 & 3 & 1 & 0 & 0 & 0 \\ 0 & 0 & 3 & 0 & 0 & 0 \\ 0 & 0 & 0 & 3 & 1 & 0 \\ 0 & 0 & 0 & 0 & 3 & 0 \\ 0 & 0 & 0 & 0 & 0 & 1 \end{bmatrix}$.

Notation

Index

Printed in the United States
by Baker & Taylor Publisher Services